T0213148

# Graduate Texts in Physics

## Graduate Texts in Physics

Graduate Texts in Physics publishes core learning/teaching material for graduate- and advanced-level undergraduate courses on topics of current and emerging fields within physics, both pure and applied. These textbooks serve students at the MS- or PhD-level and their instructors as comprehensive sources of principles, definitions, derivations, experiments and applications (as relevant) for their mastery and teaching, respectively. International in scope and relevance, the textbooks correspond to course syllabi sufficiently to serve as required reading. Their didactic style, comprehensiveness and coverage of fundamental material also make them suitable as introductions or references for scientists entering, or requiring timely knowledge of, a research field.

More information about this series at http://www.springer.com/series/8431

Bogdan Povh · Mitja Rosina

# Scattering and Structures

## Essentials and Analogies in Quantum Physics

Second Edition

 Springer

Bogdan Povh
Max-Planck-Institut für Kernphysik
Heidelberg
Germany

Mitja Rosina
Department of Physics
University of Ljubljana
Ljubljana
Slovenia

ISSN 1868-4513
Graduate Texts in Physics
ISBN 978-3-662-57202-3
DOI 10.1007/978-3-662-54515-7

ISSN 1868-4521   (electronic)

ISBN 978-3-662-54515-7   (eBook)

Printed on acid-free paper

This Springer imprint is published by Springer Nature
The registered company is Springer-Verlag GmbH Germany
The registered company address is: Heidelberger Platz 3, 14197 Berlin, Germany

# Preface to the Second Edition

Since the first edition of this book a decade ago, several new experimental and theoretical results have been obtained. While updating the new edition, we stick to our attempt to extract the essential physical content of phenomena and to illustrate them by simple *on the back of an envelope* calculations.

In particular, the Section on the Higgs boson has been put in new context and we have revised the Section on neutrino oscillations. In the Chapter on nuclear forces we have emphasized the role of pions in the nucleus. Two new Sections have been added: a Section on new allotropes of carbon such as graphene and a Section on coherent photon gas in laser.

We would like to thank Patrick Froß for his help in formatting the updates and reading the manuscript.

Heidelberg, Germany

Bogdan Povh
Mitja Rosina

# Preface to the First Edition

The initial aim of the book "*Scattering and Structures*", was to provide a revision course for German students preparing for their oral diploma and Ph.D. examinations where the student is supposed to demonstrate her or his understanding of quantum phenomena by explaining the essential physics without the ballast of the tedious details. The German edition has also been successfully used in students' seminars and in parallel with standard textbooks.

The attempt to reduce the description and explanation of complicated phenomena to the underlying ideas and formulae has motivated us to extend the framework of the book to many phenomena that seemed suited to such simplification. We hope that the book in its present format can be of interest to students and lecturers as well as to research physicists.

We have much appreciated the discussions with Bernhard Schwingenheuer (Heidelberg) on the new paragraphs of the present edition and Marcus Schwoerer (Bayreuth) for his critique of our original text on the magnetic properties of atoms and on the dispersion in crystals.

We would like to thank Martin Lavelle (Plymouth) for his excellent translation of the book and Jürgen Sawinski (Heidelberg) for his professional work in formatting it.

Heidelberg
2005

Bogdan Povh
Mitja Rosina

# Preface to the German Edition

La simplicité affectée est une imposture delicate.

La Rochefoucauld

The goal of this book may best be characterised in the words of Ernest Rutherford: *"if you can't explain a result in simple, nontechnical terms, then you don't really understand it"*. In this book *"simple, nontechnical terms"* means language that every physicist can use.

Physics may appear complicated when details cause one to lose sight of the overarching connections. Physics becomes simple when, by the application of a few basic concepts, it is possible to clarify a principle and estimate orders of magnitude. In the following, we will reveal the properties of quantum systems (elementary particles, nucleons, atoms, molecules, quantum gases, quantum liquids and stars) with the help of elementary concepts and analogies between these systems. The choice of topics corresponds to the list of themes that one of the authors (B.P.) used in Heidelberg for the oral physics diploma examination. The book is intended for preparation for the oral diploma examination and for the contemporary Ph.D. defence. Some of the chapters (e.g., 12 und 7) are, though, taken far beyond these examination levels, to make the book of interest to a wider circle of physicists. In a few cases, when we thought that current textbooks do not clearly present the latest developments in physics (e.g., Chap. 3), we have extended the size of the chapter beyond the limit that we have otherwise set ourselves.

In contrast to standard textbooks, no precise derivations are presented. Rather, we have attempted to illuminate physical connections via elementary principles (the uncertainty relation, the Pauli principle), fundamental constants (particle masses, coupling constants) and simple *on the back of an envelope* estimates. One of our models for writing the book in this style was Victor Weisskopf's lectures for summer students at CERN and his short essays *"Search for Simplicity"* published in the American Journal of Physics in 1985. The individual chapters are constructed as independent units. When we refer to other chapters, this is only to underline the analogies between different physical systems.

For each chapter, we list textbooks where the general concepts that we use and the simple formulae, which we have not derived, are to be found. All other necessary references are denoted in the text by the authors' names and are also listed at the end of each chapter.

In Chaps. 1–3 and 9, we present scattering as a method for the analysis of quantum systems. In Chaps. 4–6, we consider the construction of the simplest composites of the electromagnetic and strong interactions: atoms and hadrons. The interatomic forces that lead to the construction of molecules are treated in Chaps. 7 and 8, while the analogous force in the strong interaction, the nuclear force, is briefly discussed in Chap. 10. Degenerate systems of fermions and bosons, from quantum gases through to neutron stars, are the main theme of Chaps. 11–15. In Chap. 16, we mention some of the open questions of contemporary elementary particle physics.

It is obvious that errors can creep into any attempt to represent complex phenomena elegantly with the help of "physical intuition". We ask critical readers to point out any such slips to us. We would be happy to receive ideas for how further examples of quantum phenomena can be grasped and made plausible *on the back of an envelope*. Suggestions for how overly lengthy discussions could be shortened without any loss of clarity are also very welcome.

Special thanks for proposals improving the content, style and language of the whole book are due to Christoph Scholz (Reilingen) and Michael Treichel (Munich). The current title of the book was also suggested to us by Michael Treichel.

We received valuable criticism on the first two chapters from Paul Kienle (Munich) and on the nuclear physics chapters from Peter Brix (Heidelberg). We have discussed the treatment of chiral symmetry breaking at length with Jörg Hüfner (Heidelberg) and Thomas Walcher (Mainz). We received private tuition in phase transitions and solid state physics from Franz Wegner (Heidelberg) and Reimer Kühn (Heidelberg). Samo Fišinger (Heidelberg) helped us to formulate the section on proteins. The chapters on quantum gases and quantum liquids were produced with the help of Allard Mosk (Utrecht) and Mattias Weidemüller (Heidelberg). Claus Rolfs (Bochum) thoroughly corrected the chapter on stars. We discussed in detail the newest results from neutrino research with Stephan Schönert (Heidelberg). Ingmar Köser and Claudia Ries have taken great pains to translate the manuscript of the book into good German. Jürgen Sawinski was responsible for the layout and producing the figures.

Working with Wolf Beiglböck and Gertrud Dimler of Springer was, as ever, a pleasure.

Heidelberg                                                              Bogdan Povh
July 2002                                                               Mitja Rosina

# Prelude

The most powerful emperor of the 13th dynasty had led the Middle Kingdom to new glory. A new picture, that of a dragon, the symbol of the power of the empire, was planned to decorate his palace. He commissioned the best artist of the empire with the task of producing the picture.

After 2 years, the artist eventually appeared with his picture before the emperor. As he unrolled the canvas, the emperor glimpsed a green background with a yellow, slightly snaking line on it.

"You needed 2 years for this?" demanded the emperor in a rage. Convinced that the artist was mocking him, he let him be taken away and condemned him to death. A wise advisor of the emperor said, however: "let us, oh great emperor, personally see what the artist has done for the last 2 years." When the emperor and his advisor entered the studio of the artist, they saw over 700 pictures lined up according to the order of their production. The artist had painted a new picture each day. The first pictures showed the dragon in all possible detail. The later ones lacked more and more of the insignificant details, but the essence of the dragon was ever clearer. The last pictures were already very similar to that which the artist had brought to him. "Now I see", said the emperor, "the essence of the dragon has been perfectly represented by the artist."

The emperor pardoned the artist.

*Chinese fairy tale*

# Contents

# Chapter 1
# Photon Scattering – Form Factors

*Und so lasset auch die Farben*
*Mich nach meiner Art verkünden,*
*Ohne Wunden, ohne Narben,*
*Mit der lässlichsten der Sünden.*

Goethe

Scattering experiments are the paradigm of quantum mechanical measurements. A beam of atoms, ions, electrons or photons – to mention but a few possibilities – is generally created in an accelerator. A detector is used to find the energy of the scattered particle (or the absolute value of the momentum) and the scattering angle. From this, one calculates the momentum and energy transfer to the scattering centre and thus obtains the properties of the system under investigation.

The scattering of elementary particles (photons, leptons and quarks) off each other is distinguished by these particles not displaying any excited states, and their inter-action can be described via a fundamental coupling to exchange bosons. Scattering elementary particles off composite systems, such as atoms, nuclei or nucleons, offers the ideal method to explore their structure.

Photons are, of course, scattered off all charged particles. Since the scattering cross-section is proportional to the square of the acceleration, i.e., inversely proportional to the square of the mass of the particle, electromagnetic effects may most easily be seen in photon-electron scattering.

## 1.1 Compton Effect

The calculation of photon scattering off a free electron, Compton scattering, is a standard exercise in relativistic quantum mechanics that everyone must once endure. Here we will only treat the Klein–Nishina formula and discuss the properties of the

© Springer-Verlag GmbH Germany 2017
B. Povh and M. Rosina, *Scattering and Structures*,
Graduate Texts in Physics, DOI 10.1007/978-3-662-54515-7_1

**Fig. 1.1** Schematic
representation of both the
amplitudes (**a**) and (**b**) that
contribute to Compton
scattering at lowest order.
The electrons move in the
positive time direction, the
positrons are represented by
the negative energy electrons
which move backwards in
time

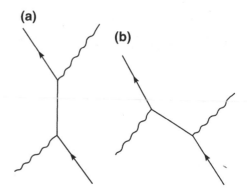

scattering in two interesting kinematic regimes. The two amplitudes that contribute
to the scattering are symbolically represented in Fig. 1.1. The famous Klein–Nishina
formula for unpolarised radiation is

$$\frac{\mathrm{d}\sigma}{\mathrm{d}\Omega'_\omega} = \frac{1}{2}r_e^2 \left(\frac{\omega'}{\omega}\right)\left(\frac{\omega'}{\omega} + \frac{\omega}{\omega'} - \sin^2\theta\right),\tag{1.1}$$

where $\hbar\omega$ and $\hbar\omega'$ are, respectively, the energies of the incoming and outgoing photons and $\theta$ is the scattering angle. The following relation links $\theta$ and the energies:

$$\cos\theta = 1 - \frac{m_e c^2}{\hbar\omega'} + \frac{m_e c^2}{\hbar\omega}.\tag{1.2}$$

Here, $r_e$ is the so-called classical electron radius, the picturesque interpretation of
which we will discuss later:

$$r_e = \frac{e^2}{4\pi\varepsilon_0 m_e c^2} = \frac{\alpha\hbar c}{m_e c^2} = \alpha\lambda_e.\tag{1.3}$$

The values of the Compton wavelength and the classical radius of the electron are
$\lambda_e = \hbar/(m_e c) = 386\,\text{fm}$ and $r_e = 2.82\,\text{fm}$. For Compton scattering with highly energetic photons ($E_\gamma \gg m_e c^2$) off electrons bound in atoms, it is a good approximation
to consider the electrons as free. In storage ring experiments, one can, however,
observe scattering off electrons that really are free, and we will briefly treat this in
Sect. 1.5.

Coherent scattering of low-energy photons off all the electrons in an atom is of
particular interest. If the atoms are bound in a crystal, the coherence of the scattering
can be extended to the entirety of the crystal.

At low energies, $E_\gamma \ll m_e c^2$, the recoil may be neglected and one can set $\omega = \omega'$.
In this approximation, the Klein–Nishina formula gives exactly the same result as
the classically calculated cross-section for Thomson scattering,

$$\frac{d\sigma}{d\Omega} = r_e^2 \frac{1 + \cos^2\theta}{2} .\tag{1.4}$$

In the following, we will ask ourselves the following: where in the amplitudes (Fig. 1.1) is the classical picture of an oscillating electron in the field of the incoming radiation, hidden? This is, anyway, the underlying picture in the derivation of the Thomson formula (1.4).

## 1.2 Thomson Scattering

### 1.2.1 Classical Derivation

Let us first consider the scattering of linearly polarised light off an electron in an atom (Fig. 1.2). Neglecting the recoil, the electron moves in the electric field $\mathbf{E}_0 e^{i\omega t}$ of the incoming light wave, and its acceleration is

$$\mathbf{a} = \mathbf{E}_0 \frac{e}{m} e^{i\omega t} .\tag{1.5}$$

The accelerated charge radiates. For those waves that spread out perpendicularly to the induced dipole, the electric field strength in the radiation zone is proportional to the product of the acceleration and the charge,

**Fig. 1.2** Coherent photon scattering off an atom. The polarisation vector is (**a**) in the plane ($\vartheta = \pi/2 - \theta$), (**b**) orthogonal to the plane ($\vartheta = \pi/2$)

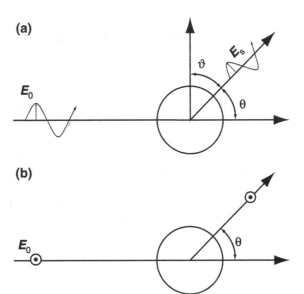

$$E_s(t, r, \vartheta = \pi/2) = \frac{1}{4\pi\varepsilon_0} \frac{e^2}{mc^2} E_0 \frac{e^{i(\omega t - kr)}}{r}, \tag{1.6}$$

where the factor $1/(4\pi\varepsilon_0)$ ensures the correct units and the $1/r$-dependence preserves energy conservation because $\int E_s^2 r^2 d\Omega$ must be independent of $r$.

The amplitude of the electric field strength of the radiation in any direction where $\vartheta \neq \pi/2$ is reduced. The reduction factor is $\sin\vartheta$, where $\vartheta$ is measured from the polarisation direction of the incoming wave. This factor yields the projection of the polarisation vector of the incoming radiation with respect to the polarisation direction of the radiation field (Fig. 1.2).

The energy density,

$$\frac{1}{2}(\varepsilon_0 E_s^2 + \mu_0 B_s^2) = \varepsilon_0 E_s^2, \tag{1.7}$$

multiplied by $c$ yields the energy flux. The energy flux scattered into the solid angle $d\Omega$ is thus found to be

$$\begin{aligned} c\varepsilon_0 E_s^2 r^2 d\Omega &= \frac{c\varepsilon_0 E_0^2}{(4\pi\varepsilon_0)^2} \left(\frac{e^2}{mc^2}\right)^2 \sin^2\vartheta d\Omega \\ &= c\varepsilon_0 E_0^2 r_e^2 \sin^2\vartheta d\Omega. \end{aligned} \tag{1.8}$$

The so-called classical electron radius is a measure of the acceleration of an electron in an electric field. It has nothing to do with the geometrical extension of the electron.

Its historical denotation as a radius came from the relation

$$mc^2 = \frac{e^2}{4\pi\varepsilon_0 r_e}. \tag{1.9}$$

The electrostatic energy of a sphere with radius $r_e$ and charge $e$ is in the classical picture related to the electron mass. The appearance of the radius $r_e$ in electrodynamics has a plausible explanation. If two electrons come together up to a separation $r_e$, then the potential energy is so great that an $e^+e^-$ pair is created; the concept of a single electron thus loses its meaning. For nonpolarised light, one measures the angle $\theta$ from the beam direction (Fig. 1.2). The total intensity of the scattered light is found from incoherently averaging the contributions (1.8) of the two orthogonal polarisation states. In atoms with $Z$ electrons and for wavelengths large in comparison with the atomic radius, the electrons oscillate with the same phase and the contributions to the scattering off the individual electrons are added coherently,

$$c\varepsilon_0 E_s^2 d\Omega = c\varepsilon_0 E_0^2 Z^2 r_e^2 \frac{1 + \cos^2\theta}{2} d\Omega. \tag{1.10}$$

The photon flux, i.e., the number of photons that hit the target per unit area per second, is $\Phi_0 = c\varepsilon_0 E_0^2/(\hbar\omega)$. The number of photons scattered into the solid angle $d\Omega$ is found from

$$\Phi_s d\Omega = \Phi_0 Z^2 r_e^2 \frac{1 + \cos^2\theta}{2} d\Omega, \tag{1.11}$$

from which the differential cross-section

$$\frac{d\sigma}{d\Omega} = Z^2 r_e^2 \frac{1 + \cos^2\theta}{2} \tag{1.12}$$

may be deduced.

## 1.2.2 Quantum Mechanical Derivation

The same result as above can be very simply derived quantum mechanically at low energies. Because we may calculate nonrelativistically, the interaction between photons and electrons is given by the following Hamiltonian:

$$\frac{(\mathbf{p} - e\mathbf{A})^2}{2m_e} = \frac{\mathbf{p}^2}{2m_e} - \frac{e\mathbf{A} \cdot \mathbf{p}}{m_e} + \frac{e^2\mathbf{A}^2}{2m_e}. \tag{1.13}$$

The first term corresponds to the kinetic energy of the electron and the rest to the perturbation. In Fig. 1.3 the amplitudes that are proportional to $\alpha$ are represented diagrammatically. Amplitudes (a) and (b) have the form

$$M \sim \frac{e\langle\mathbf{A} \cdot \mathbf{p}\rangle}{m_e} \frac{1}{\Delta E} \frac{e\langle\mathbf{A} \cdot \mathbf{p}\rangle}{m_e}. \tag{1.14}$$

When one explicitly writes out both amplitudes, one can easily convince oneself that they have opposite signs and cancel each other as $\omega' \to \omega$. It is easy to see that both amplitudes have opposite signs because the amplitudes have (a) $\Delta E = +\hbar\omega$ and (b) $\Delta E = -\hbar\omega$. Furthermore, for energies $\hbar\omega \ll m_e c^2$, the amplitudes (a) and (b) are anyway small compared with the amplitude (c). The first two contain two separate

**Fig. 1.3** The amplitudes that contribute to Compton scattering in the nonrelativistic limit

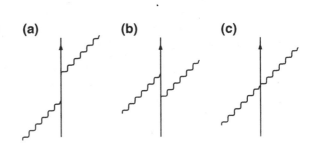

vertices and so have a factor of $m_e^2$ in the denominator, while in amplitude (c), there is only one power of the electron mass.

Considered superficially, one might think that the amplitude (c) is a limiting case of (a) and (b); this is, however, not the case, as we will see in the following: If we want to calculate the amplitude (c), we have to quantise the electromagnetic field, $\mathbf{A}$. When a photon with polarisation $\varepsilon$ is created or annihilated, the expectation value of $\mathbf{A}$ is given by $(\hbar/\sqrt{2\varepsilon_0\hbar\omega})\varepsilon$. To make this "photon normalisation" plausible, we consider an electromagnetic eigenmode (with periodic boundary conditions) in a normalisation volume: $E/\mathcal{V} = \varepsilon_0\mathbf{E}^2/2 + \mathbf{B}^2/2\mu_0 = \varepsilon_0|d\mathbf{A}/dt|^2/2 + |\nabla \times \mathbf{A}|^2/2\mu_0 = \varepsilon_0[(\omega A)^2 + c^2(kA)^2]/2 = \varepsilon_0\omega^2 A^2 = \hbar\omega/2$. We have expressed the electric and magnetic fields in terms of $A$; both fields give the same contribution. The amplitude (c) is then given as $\omega' \to \omega$ by

$$M = 2\frac{e^2}{2m_e}\frac{\varepsilon_i}{\sqrt{\varepsilon_0}\sqrt{2\hbar\omega}}\frac{\hbar}{\sqrt{\varepsilon_0}\sqrt{2\hbar\omega}} \cdot \frac{\varepsilon_f}{\sqrt{\varepsilon_0}\sqrt{2\hbar\omega}}\frac{\hbar}{\sqrt{\varepsilon_0}\sqrt{2\hbar\omega}} = \frac{2\pi r_e(\hbar c)^2}{\hbar\omega}\varepsilon_i \cdot \varepsilon_f, \tag{1.15}$$

where $\varepsilon_i$ and $\varepsilon_f$ are, respectively, the polarisation vectors of the incoming and outgoing photons. Their scalar product is either 1 (Fig. 1.2b) or $\cos\theta$ (Fig. 1.2a). The cross-section obtained in this way for unpolarised radiation off $Z$ electrons is then

$$\frac{d\sigma}{d\Omega} = \frac{2\pi}{\hbar}Z^2\overline{|M|^2}\frac{(\hbar\omega/c)^2}{(2\pi\hbar)^3c^2} = Z^2 r_e^2\frac{1 + \cos^2\theta}{2}, \tag{1.16}$$

which is identical to the classically derived equation (1.12).

## 1.2.3 Quantum Mechanical Interpretation of $r_e$

Superficially considered, it sounds surprising that the Dirac equation yields the same result (1.12) in the nonrelativistic limit, despite the corresponding amplitude depicted in Fig. 1.3c not explicitly appearing. The explanation is as follows: in the relativistic case, the propagator in the amplitudes (a) and (b) of Fig. 1.1 also contains positrons. In Fig. 1.4, the positrons are explicitly shown in the two diagrams labeled by (c). While the amplitudes (a) and (b) vanish for small velocities due to the current coupling $\sqrt{\alpha}p/m_ec$, the photon coupling to the electron–positron pair is $\sqrt{\alpha}$. In the case of pair creation, the intermediate state involves two additional electron masses and the propagator is proportional to $1/2m_e$. It follows from this that the amplitudes labeled (c) in Fig. 1.4 are proportional to $\langle e^2\mathbf{A}^2/2m_e\rangle$.

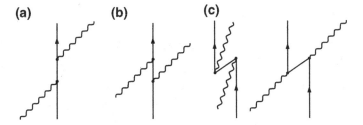

**Fig. 1.4** The contribution of electron–positron pairs (**c**) to Compton scattering

It is well worth stressing that the classical oscillations of an electron in an electromagnetic field in the relativistic case correspond to the coupling of the photon to electron–positron pair fluctuations in the vacuum. This means that Thomson scattering in the relativistic calculation results from a sum of the contributions of the small components of the Dirac wave function.

The classical electron radius also acquires a new interpretation: Thomson scattering is proportional to

$$r_e^2 = \alpha \cdot \alpha \cdot \lambdabar_e^2, \tag{1.17}$$

i.e., proportional to the probability that one finds an electron–positron pair inside its range ($\propto \lambdabar_e^2$) and proportional to the probability that this electron–positron pair interacts with the incoming ($\alpha$) and the outgoing photon ($\alpha$).

## 1.3 Form Factor

The scattering of elementary particles off composite systems is the best method to measure their extension.

### 1.3.1 Geometrical Interpretation of the Form Factor

When the wavelengths of X-rays are comparable with the extension of an atom, one has to take into account the phases of the waves that are scattered off different regions of the atom. In Fig. 1.5, the planes orthogonal to the momentum transfer vector **q** are sketched. They are denoted by dashed lines and are orthogonal to the plane of the page.

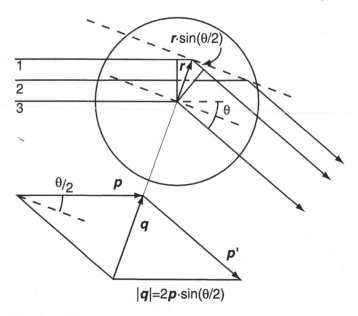

**Fig. 1.5** Diffraction of X-rays off an atom

All beams that are scattered in the same plane (beams 1, 2) have the same phase and their amplitudes add together completely. It is therefore sufficient to only consider one beam in every additional plane and determine the phase differences relative to the plane that passes through the middle of the sphere (beam 3). The path length difference between beam 1 and beam 3 is $\Delta = 2r\sin(\theta/2)$ and the phase is

$$2\pi\Delta/\lambda = 2pr\sin(\theta/2)/\hbar = \mathbf{q}\mathbf{r}/\hbar, \tag{1.18}$$

where $\lambda = 2\pi\hbar/p$. The amplitude of the radiation elastically scattered through the angle $\theta$ is then reduced by the factor

$$F(q^2) = \int \rho(\mathbf{r})e^{i\mathbf{q}\mathbf{r}/\hbar}\mathrm{d}^3r. \tag{1.19}$$

We call this factor the form factor. It is the Fourier transform of the charge density $\rho(\mathbf{r})$ of the atom. The differential cross-section for the scattering of X-rays off atoms is thus

$$\frac{\mathrm{d}\sigma}{\mathrm{d}\Omega} = Z^2 r_e^2 F^2(q^2)\frac{1 + \cos^2\theta}{2}. \tag{1.20}$$

If we write the expectation value of the square of the atomic radius as $\langle r^2\rangle$ and expand (1.19) in $q^2$ around $q^2 = 0$, then we obtain

**Fig. 1.6** Electron density distribution in NaCl crystal. The *numbers* show the relative electron density

$$F(q^2) = 1 - \frac{q^2}{2\hbar^2}\overline{\cos^2\theta} \int r^2\rho(r)4\pi r^2 dr + \dots$$

$$= 1 - \frac{\langle r^2\rangle}{6\hbar^2}q^2 + \dots , \tag{1.21}$$

where the average over $\cos^2\theta$, as is well known, is 1/3.

Atomic form factors have been found from X-ray diffraction off crystals. In Fig. 1.6, the experimentally determined electron densities of $Na^+$ and $Cl^-$ ions in NaCl crystal are depicted. These densities roughly correspond to those of the noble gases neon and argon. To extract the form factors, one has to divide the density distributions by $Z^2$ for both the ions. Such normalised density distributions in noble gases have almost identical extensions and are described to a very good approximation by an exponential function (Fig. 5.3), the Fourier transform of which is

$$F(q^2) \approx \frac{1}{[1 + (qa)^2]^2}, \tag{1.22}$$

where $a^2 = \langle r^2\rangle/(12\hbar^2)$. The mean square radii of both ions are comparable: $\sqrt{\langle r^2\rangle} \approx 0.13\,\text{nm}$.

In Fig. 1.6, the relative electron densities are shown and the $Cl^-$ ion seems larger than the $Na^+$ ion.

### 1.3.2   Dynamical Interpretation of the Form Factor

Let us attempt to give a dynamical interpretation of the form factor. The extension of the atom is linked to the binding energy of the electron in the Coulomb field by the uncertainty relation. In place of the binding energy, we introduce the idea of the typical excitation of the system, which we denote by $D$. In the case of the oscillator potential, $D$ is the separation of the excited states, while, for atoms, $D$ is of the order of magnitude of the binding energy. The expectation value of $\langle r^2 \rangle$ can then be approximately replaced by $D$,

$$\langle r^2 \rangle = f \frac{\hbar^2}{\langle p^2 \rangle} = f \frac{\hbar^2}{2 m_e D} \, . \tag{1.23}$$

The value of $f$ depends on the specific potential but is of the order of magnitude of 1. The form factor (1.21) can then be expressed in terms of the typical excitation of the system $D$ (1.23),

$$F(q^2) = 1 - \frac{f}{12 m_e D} q^2 + \dots \, . \tag{1.24}$$

For increasing momentum transfer, the recoil energy will eventually suffice to excite the electron into a higher energy state or into the continuum. The probability that the system remains in the ground state after the scattering decreases rapidly for

$$\frac{q^2}{2 m_e} \geq D \, . \tag{1.25}$$

## 1.4   Recoilless Scattering Off Crystals

Because the Nobel prize has twice been awarded (von Laue 1920, Mössbauer 1957) for the discovery of recoilless X-ray scattering off crystals and for gamma emission in crystals, we want here to derive *on the back of an envelope* the probability that the scattering takes place off the entire crystal.

Consider atoms bound in a crystal where the interatomic potential has the form of a harmonic oscillator. The typical excitation is $D = \hbar\omega$. Let us consider an atom in the ground state for which the wavefunction is

$$\psi_0(r) = \left( \frac{M\omega}{\hbar\pi} \right)^{3/4} e^{-M\omega r^2/(2\hbar)} \, . \tag{1.26}$$

Immediately after the recoil, the wavefunction has not had time to change its spatial form; however, the momentum received can be seen in the phase factor $\exp(iqr/\hbar)$,

$$\psi_0(r) \rightarrow \psi'(r) = e^{iqr/\hbar}\psi_0(r). \tag{1.27}$$

The probability that the atom remains in the ground state is the square of the overlap between the new wave function, $\psi'$, and the ground-state wave function, $\psi_0$,

$$P(0, 0) = \left|\langle\psi_0|e^{iqr/\hbar}\psi_0\rangle\right|^2 = \left|\int \psi_0^* e^{iqr/\hbar}\psi_0 d^3r\right|^2 = e^{-q^2/(2M\hbar\omega)}. \tag{1.28}$$

Now we must define the typical excitation of the crystal $D$ or $\hbar\omega$. In the Debye model of the crystal, $D \approx \frac{2}{3}k\Theta$, where $\Theta$ is the Debye temperature. When we substitute this value for $D$ into (1.28), we obtain

$$P(0, 0)_{DW} = e^{-3q^2/(4Mk\Theta)}. \tag{1.29}$$

Equation (1.29) is the simplified form of the Debye–Waller factor for $T = 0\,K$. It gives the probability of coherent scattering off crystals and also for the recoilless emission of gamma rays from crystalline sources (the Mössbauer effect). To underline the complementarity of the dynamical and geometrical interpretations of the form factor, let us again repeat that the Debye–Waller factor is the form factor of an atom bound in a crystal.

## 1.5 Photon Scattering Off Free Electrons

Photon scattering (or Compton scattering) off free electrons may be easily performed at electron storage rings and has many applications in accelerator physics. At DESY, for example, a laser beam with $\hbar\omega = 2.415\,eV$ hits $27.570\,GeV$ electrons. The backward scattered photons are in the energy spectrum of high energy gamma rays, with an energy of $13.92\,GeV$ (Fig. 1.7).

The energy of the backward scattered photon can be easily estimated when one equates the relativistically invariant quantity $s$, the square of the centre of mass energy, before the scattering with its value after the scattering. Before the scattering,

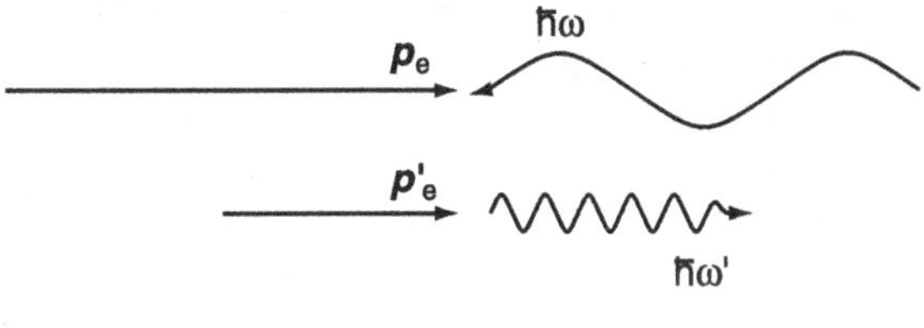

**Fig. 1.7** Scattering of laser light off a high energy electron

$$s = (E_e + E_\gamma)^2 - (p_e c - E_\gamma)^2$$
$$= m_e^2 c^4 + 2E_\gamma (E_e + p_e c) \qquad (1.30)$$
$$\approx m_e^2 c^4 + 4E_\gamma E_e ,$$

where we assumed $E_e \approx p_e c$, and after the scattering, we have

$$s' = (E'_e + E'_\gamma)^2 - (p'_e c + E'_\gamma)^2$$
$$= m_e^2 c^4 + 2E'_\gamma (E'_e - p'_e c) \qquad (1.31)$$
$$\approx m_e^2 c^4 + E'_\gamma \frac{m_e^2 c^4}{E'_e} .$$

The final step in (1.31) is obtained when one multiplies $s'$ by $(E'_e + p'_e c)/2E'_e$ and then assumes $E'_e \approx p'_e c$.

Making use of conservation of energy, $E'_e \approx E_e - E'_\gamma$, comparison of the two expressions for $s$ yields

$$E'_\gamma = 4E_\gamma E_e \frac{E_e - E'_\gamma}{m_e^2 c^4}, \qquad (1.32)$$

which leads to the result

$$E'_\gamma = \frac{E_e}{1 + m_e^2 c^4/(4E_\gamma E_e)} = E_e \cdot \frac{4E_\gamma E_e}{s}. \qquad (1.33)$$

For the energies mentioned above, $m_e^2 c^4/4E_\gamma E_e = 0.98$ and $E'_\gamma \approx E'_e \approx E_e/2$. From this, it follows that the centre of mass energy, $\sqrt{s} \approx \sqrt{2} m_e c^2$. This value contains the rest energy $m_e c^2$, so the kinetic energy is only a fraction of the total energy. Therefore, we may estimate the cross-section nonrelativistically, and we can take the Thomson value

$$\sigma = \frac{8}{3} \pi r_e^2. \qquad (1.34)$$

The exact calculation of the Klein–Nishina cross-section integrated over $4\pi$ yields, for the example treated above, a result that is smaller by a factor of 0.81. For centre of mass energies below twice the electron mass, $pc \leq m_e c^2$, the Thomson cross-section is a good estimate.

Compton scattering is generally taken to refer to quasi-elastic photon scattering off an electron in an atom. For lower energy and angular resolution of the measurement, it suffices to calculate the kinematics of the scattering off a static electron. The quality of contemporary detectors is, though, sufficient to observe the influence of atomic, molecular or solid state effects on the kinematics of the scattered particle.

# Literature

R.P. Feynman, *Quantum Electrodynamics* (Benjamin, New York, 1962)

R.P. Feynman, R.B. Leighton, M. Sands, *The Feynman Lectures on Physics II* (Addison-Wesley, Reading, 1964)

# Chapter 2
# Lepton Scattering – Nucleon Radius

*J.J. Thomson got the Nobel prize for demonstrating that the electron is a particle. George Thomson, his son, got the Nobel prize for demonstrating that the electron is a wave. For me the electron is simply a second quantized relativistic field operator.*
Physics Colloquium, Heidelberg 2001 Cecilia Jarlskog

Electron scattering off protons yielded the first indications for the finite size of the proton (Hofstadter 1957) and, later, the experimental evidence (Friedman, Kendall, Taylor 1967) for the modern parton model of the nucleon.

In recent decades, neutrino experiments have become fashionable. One of the goals of these experiments is to determine the masses of neutrinos by observing oscillations between different families of neutrinos. Such oscillations have in fact been observed, demonstrating that neutrinos are not massless. The neutrinos are observed in detectors that can recognise elastic scattering off an electron through its recoil or can identify elastic scattering with charge exchange off a quark. Another goal is to study nucleon properties via weak interaction.

In this chapter, we will clarify the analogies between electron-quark, neutrino-electron and neutrino-quark scattering.

## 2.1 Electron-Quark Scattering

The symbols for the quantities that describe this scattering process are defined in Fig. 2.1.

In scattering, the square of the four-momentum transfer is negative ($q^2 < 0$), one therefore rather uses the variable $Q^2 = -q^2$. The virtual photon has an invariant mass, $M_\gamma = Q/c$, and energy $\nu$. In the laboratory frame, the square of the photon

© Springer-Verlag GmbH Germany 2017
B. Povh and M. Rosina, *Scattering and Structures*,
Graduate Texts in Physics, DOI 10.1007/978-3-662-54515-7_2

**Fig. 2.1** Electron-quark scattering: $e, e', q$ and $q'$ are four-vectors, $Q^2$ is the negative square of the four-momentum transfer and $\nu$ is the energy transfer

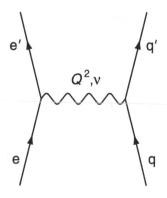

mass (multiplied by $c^4$) is $(Qc)^2 = 4EE' \sin^2(\theta/2)$ and the energy of the photon is $\nu = E - E'$. Here, $\theta$ is the scattering angle of the electron.


## 2.1.1 Mott Scattering

The scattering of an electron off a spin zero, charged particle is called Mott scattering. The photon propagator is well known to be proportional to $1/Q^2$. Various paths lead to this result. Here we will offer an alternative derivation, which will make clear why the photon propagator depends on the square of the momentum transfer and why the range of a virtual particle decreases exponentially with respect to its mass.

Two amplitudes contribute to the scattering. The electron can emit a photon, as can the quark. To determine the matrix elements, we have to find the virtuality of both photons. For real photons, the relation $\hbar\omega = |\mathbf{q}|c$ holds; therefore, the virtuality of the one photon is $\Delta E_1 = \hbar\omega - |\mathbf{q}|c$, while the virtuality of the other is $\Delta E_2 = \hbar\omega' - |\mathbf{q}|c$. The coupling constant at the vertices is, of course, the charge. To avoid the unpleasant factor $\varepsilon_0$ of the SI system, we write the photon-electron coupling constant as $e/\sqrt{\varepsilon_0} = \sqrt{4\pi\alpha\hbar c}$ and the photon-quark coupling constant as $z_q e/\sqrt{\varepsilon_0} = z_q\sqrt{4\pi\alpha\hbar c}$. As our normalisation of the photon wave function, we choose $\hbar c/\sqrt{2|\mathbf{q}|c}$.

To make this normalisation plausible, let us consider an electromagnetic eigenmode with the periodic boundary conditions in a normalisation volume: $E/V = \varepsilon_0 \mathbf{E}^2/2 + \mathbf{B}^2/2\mu_0 = \hbar\omega/2$. Because the electric and magnetic contributions are equal, we may express the energy through the electric potential, $\phi$. The electric field strength is proportional to the potential $\phi$, so $\varepsilon_0 \mathbf{E}^2 = \varepsilon_0 (k\phi)^2 = \hbar\omega/2$. The interaction of the electron with the electric field is then $H' = e\phi = (e/k)\sqrt{\hbar\omega/2\varepsilon_0} = (e/\sqrt{\varepsilon_0})\,(\hbar c)/\sqrt{2\hbar\omega}$ (see also (1.15)).

**Fig. 2.2** The two contributions to the scattering amplitude

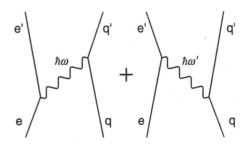

The scattering amplitude is then (cf. Fig. 2.2)

$$M = 4\pi\hbar c \frac{\sqrt{\alpha}\hbar c}{\sqrt{2|\mathbf{q}|c}} \left( \frac{1}{\hbar\omega - |\mathbf{q}|c} + \frac{1}{\hbar\omega' - |\mathbf{q}|c} \right) z_q \frac{\sqrt{\alpha}\hbar c}{\sqrt{2|\mathbf{q}|c}}. \tag{2.1}$$

The unit volume, which we employed for the normalisation, cancels out in the final result and we have therefore not explicitly written it in (2.1).

For the energy transfer from the quark to the electron, we have $\omega' = -\omega$. We can thus, in the above expression, rename the integration variable $\omega'$. The scattering amplitude can then be written as

$$M = -\frac{4\pi z_q \alpha(\hbar c)^3}{(\hbar\omega)^2 - (\mathbf{q}c)^2} = -\frac{4\pi z_q \alpha(\hbar c)^3}{q^2 c^2}, \tag{2.2}$$

where we recognise the well known form of the photon propagator. Because, in the scattering, the four-momentum transfer is such that $q^2 < 0$, one uses the variable $Q^2 = -q^2$.

It is clear from the above discussion why it is the square of the virtuality of the exchange particle that appears in the denominator of the propagator for the photon and for all other boson propagators: the two amplitudes of the exchange bosons (one from left to right and one from right to left) represent a symmetric state. The sum of these amplitudes is inversely proportional to the square of the momentum transfer.

If relativistic electrons are scattered in a Coulomb field, the helicity

$$h = \frac{\mathbf{s} \cdot \mathbf{p}}{|\mathbf{s}| \cdot |\mathbf{p}|} \tag{2.3}$$

is conserved. Assume the electron has spin in the direction of the beam. For a scattering angle $\theta$, helicity conservation leads to an additional factor $\cos(\theta/2)$ (Fig. 2.3) in the amplitude. It follows that the scattering vanishes at 180°, which is made plausible in Fig. 2.3.

**Fig. 2.3** Helicity is
conserved in the limit
$v/c \to 1$. This is not
possible for scattering
through 180° off a spin zero
target due to conservation of
angular momentum

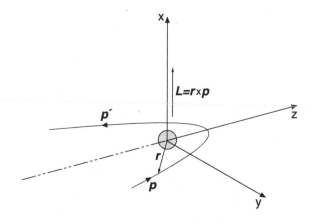

We can now write down the scattering formula for relativistic electrons off a spin zero charged quark. The number of scatterings per unit time $W$ (the Golden Rule) is

$$c\sigma = \frac{2\pi}{\hbar}|M^2|\frac{dn}{dE} \ . \tag{2.4}$$

As with the matrix element, we have neglected the normalisation volume also here. The Mott scattering formula for electrons off spinless quarks with charge $z_q e$ into a solid angle $d\Omega$ and energy $E'$ is then explicitly

$$\frac{d\sigma(eq \to eq)}{dE'd\Omega} = \frac{4z_q^2\alpha^2 E'^2(\hbar c)^2}{Q^4 c^4}\cos^2\frac{\theta}{2}\delta\left(\nu - \frac{Q^2}{2m}\right) \ . \tag{2.5}$$

Because of the recoil, $E \neq E'$. The delta function, which comes into play because of the phase space $dn/dE$, ensures the correct relation between $Q^2$ and $\nu = E - E'$ for elastic scattering, as we wish to briefly explain.

The quark receives energy $\nu$ and three momentum $\mathbf{q}$ from the photon. The invariant mass $m$ of the quark after the scattering is

$$(\nu + mc^2)^2 - (\mathbf{q}c)^2 = (mc^2)^2 \ , \tag{2.6}$$

which implies that

$$\nu^2 + 2mc^2\nu + (mc^2)^2 - (\mathbf{q}c)^2 = (mc^2)^2 \ . \tag{2.7}$$

Our definition of the four-momentum transfer implies

$$-(Qc)^2 = (qc)^2 = \nu^2 - (\mathbf{q}c)^2 \tag{2.8}$$

and

$$Q^2 = 2m\nu \,. \tag{2.9}$$

The usual expression for the Mott cross-section is found by integrating over $E'$ at constant angle $\theta$ (and equivalently at constant $Q^2/E'$). The integral over the delta function is performed using the fact that $\delta(ax) = \delta(x)/a$.

$$
\begin{aligned}
&\int \delta\left(E - E' - \left(\frac{Q^2}{2mE'}\right)E'\right) \mathrm{d}E' \\
&= \int \frac{\delta(E' - E/(1 + Q^2/2mE'))}{1 + Q^2/2mE'} \mathrm{d}E' \\
&= \frac{1}{1 + Q^2/2mE'} = \frac{1}{1 + (E - E')/E'} = \frac{E'}{E}\,.
\end{aligned}
\tag{2.10}
$$

In the last step, we have used (2.9) to express the result in terms of $E'$. The differential cross-section for the Mott scattering in the usual form and with $z_q = 1$ is then

$$\frac{\mathrm{d}\sigma_{\mathrm{Mott}}}{\mathrm{d}\Omega} = \frac{4\alpha^2 E'^2(\hbar c)^2}{Q^4 c^4} \frac{E'}{E} \cos^2\frac{\theta}{2}\,. \tag{2.11}$$

## 2.1.2   Inclusion of Quark Spin

Quarks, however, have spin $s = 1/2$ and charge $z_q e$. Accordingly, they have a magnetic moment. In scattering of charged particles with magnetic moments, a spin flip takes place. This contribution is proportional to the four-momentum transfer and $\sin^2(\theta/2)$

$$
\begin{aligned}
\frac{\mathrm{d}\sigma(\mathrm{eq} \to \mathrm{eq})}{\mathrm{d}E'\mathrm{d}\Omega} = {}&\frac{4z_q^2\alpha^2 E'^2(\hbar c)^2}{Q^4 c^4} \delta\left(\nu - \frac{Q^2}{2m}\right) \\
&\times \left(\cos^2\frac{\theta}{2} + 2\frac{Q^2}{4m^2 c^2}\sin^2\frac{\theta}{2}\right).
\end{aligned}
\tag{2.12}
$$

Using (2.11), we can write (2.12) in a more compact fashion for electron-quark scattering,

$$\frac{\mathrm{d}\sigma(\mathrm{eq} \to \mathrm{eq})}{\mathrm{d}\Omega} = \frac{\mathrm{d}\sigma_{\mathrm{Mott}}}{\mathrm{d}\Omega} z_q^2\left(1 + 2\tau\tan^2\frac{\theta}{2}\right), \tag{2.13}$$

where

$$\tau = \frac{Q^2}{4m^2 c^2}\,. \tag{2.14}$$

## 2.2  Electron-Nucleon Scattering

Because the charges in the nucleon are carried by quarks, it is justified to describe
elastic electron-nucleon scattering at energies below $200\,\text{MeV}$ ($\lambda = \hbar/p \approx 1\,\text{fm}$)
as coherent scattering off the quarks. To be able to apply the formula (2.13) to
electron-nucleon scattering, we must take account of the following: the nucleon
is a composite system with a finite extension and a magnetic moment that is not
that of a Dirac particle ($g \neq 2$). We describe the finite extension through a form
factor for both the electric charge distribution and another for the distribution of the
magnetisation. The anomalous magnetic moment is not only important in magnetic
scattering but also in electric scattering – through electric fields induced via the
anomalous magnetic moment. These corrections are usually parameterised through
the so-called Rosenbluth formula,

$$\frac{d\sigma}{d\Omega} = \frac{d\sigma_{\text{Mott}}}{d\Omega} \left[ \frac{G_E^2(Q^2) + \tau G_M^2(Q^2)}{1 + \tau} + 2\tau G_M^2(Q^2) \tan^2 \frac{\theta}{2} \right]. \qquad (2.15)$$

Here, $G_E^2(Q^2)$ and $G_M^2(Q^2)$ are the electric and magnetic form factors, which
depend on $Q^2$. They are so normalised that, as $Q^2 \to 0$, they yield the total charge
and magnetic moment in nuclear magneton units. Thus, for the proton, $G_E^p(Q^2 =
0) = 1$ and $G_M^p(Q^2 = 0) = 2.79$, while for the neutron, $G_E^n(Q^2 = 0) = 0$ and $G_M^n
(Q^2 = 0) = -1.91$.

In contrast to the notation $F$ for the form factor of the Dirac particles, in the form
factors $G$ the anomalous magnetic moment is included.

### 2.2.1  Nucleon Radius

The expectation value of the square of the charge radius is given by (1.21)

$$\langle r^2 \rangle = -6\hbar^2 \left( \frac{dG_E^2}{dQ^2} \right)_{Q^2=0}, \qquad (2.16)$$

and the corresponding expression for the magnetic radius contains the derivative of
$G_M$ with respect to $Q^2$. The charge radius of the proton and the magnetisation radii
of the proton and the neutron are roughly the same size. The value of $\sqrt{\langle r^2 \rangle}$ lies
between $0.81\,\text{fm}$ and $0.89\,\text{fm}$, depending on in which $Q^2$ domain the derivative of
the form factor is calculated.

### 2.2.2 Nucleon Form Factor

Both the radius and the proton form factor are experimentally known up to $Q^2 \approx$ 20 GeV$^2$. For $Q^2 \geq 0.2$ GeV$^2$, the form factor can be described through a so-called dipole fit,

$$G_E(Q^2) = \left[1 + \frac{Q^2}{0.71(\text{GeV}/c)^2}\right]^{-2} = 1 - \frac{Q^2}{0.36(\text{GeV}/c)^2} + \cdots . \qquad (2.17)$$

Let us now try to relate the form factor (2.17) to a typical nucleon excitation (1.24). The first excited state of the nucleon with negative parity lies $\approx 0.6$ GeV above the ground state. We identify this energy with the typical nucleon excitation, $D$. We replace the electron mass in equation (1.24) with the mass of the constituent quark, $m_q = 0.35$ GeV. Thus the form factor to a first approximation is

$$F(Q^2) = 1 - \frac{Q^2}{2m_q D} = 1 - \frac{Q^2}{0.42(\text{GeV}/c)^2} + \cdots , \qquad (2.18)$$

in good agreement with (2.17) – which is a further demonstration that the extension of a quantum object and its excitations are closely related through the uncertainty relation.

When one interprets the form factor (2.17) as the Fourier transform of the charge distribution, then the latter has the form

$$\rho(r) = \rho(0)e^{-2r/a_0^p}, \qquad (2.19)$$

where $a_0^p = 0.47$ fm. Interestingly, the radial dependence of the charge distribution in the proton displays the same exponential form as that of the hydrogen atom. Were one able to treat the proton as a nonrelativistic system, then the static gluon field in the proton would have a $1/r$ dependence and the Bohr radius $a_0$ of the proton would have the value $a_0^p \approx 0.5$ fm.

This is not a great surprise. The typical excitation energy in a hydrogen atom is $D_H \approx 10$ eV, and in a proton, it is $D_p \approx 0.6$ GeV. Thus we expect

$$\frac{a_0^p}{a_0} \approx \frac{\alpha_s/D_p}{\alpha/D_H} \approx 10^{-5} \qquad (2.20)$$

We have here assumed that the ratio between the strong and the electromagnetic coupling is $\alpha_s/\alpha \approx 100$. The surprise is, however, the charge distribution of the proton that corresponds to the quarks in a simple Coulomb field. This is not properly described in any model. A $1/r$ potential is plausible for small

distances between the quarks; however, for the larger separations corresponding to small $Q^2$, one would expect the influence of confinement to make itself
visible. In practice, the charges inside the proton are carried by constituent
quarks and by mesons (quark antiquark pairs), which do not feel confinement,
and are responsible for the charge distribution on the periphery. A theoretical
description of the confinement phenomenon still does not exist.

## 2.3 Neutrino-Electron Scattering

Neutrino scattering off the electron is the weak interaction's version of Rutherford or
Mott scattering: instead of photons, $Z^0$ bosons are exchanged (Fig. 2.4). The principal
difference from the electromagnetic case is due to the large mass of the $Z^0$ exchange
boson. This also does not couple to all lepton pairs with the same strength. We will
therefore use an effective weak coupling constant, $\tilde{\alpha}_Z = f\alpha_Z$. The factor $f$ is of
the order of magnitude 1 and we will discuss it in more detail in Chap. 16 about the
electro-weak interaction.

The formula analogous to (2.11) for $\nu e \to \nu' e'$ scattering is then

$$\frac{d\sigma(\nu e \to \nu' e')}{d\Omega} = \frac{4(\tilde{\alpha}_Z \hbar c)^2 E'^2}{\{(\hbar\omega)^2 - [(\mathbf{q}c)^2 + (m_Z c^2)^2]\}^2} \frac{E'}{E} \cos^2 \frac{\theta}{2} f^2(\theta), \qquad (2.21)$$

where we denote by $f^2(\theta)$ the spin dependence of the angular distribution.

For lower energies (below about 10 GeV), we can neglect $\hbar\omega$ and $\mathbf{q}c$ compared
to the mass $M_Z c^2 = 91$ GeV. One recognises that, in contrast to Rutherford or Mott
scattering, the forward divergence $1/\sin^4(\theta/2)$ is not present.

Let us now make an *on the back of an envelope* estimation of the total neutrino-
electron cross-section. Let us go to the centre of mass frame ($E_{cm} = E'_{cm}$) and,
because the angular dependence in this frame is not great, we replace the solid angle
integral by $4\pi$. The ratio of the weak and electromagnetic couplings is $\alpha_W/\alpha \approx 4$,

**Fig. 2.4** Neutrino scattering
off the electron without
charge exchange

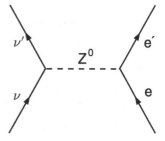

but for rough estimates, we employ $\tilde{\alpha}_W \approx \tilde{\alpha}_Z \approx \alpha$. We so obtain for the integrated cross-section

$$\sigma(\nu e \rightarrow \nu'e') \approx 4\pi \frac{4(\alpha\hbar c)^2 E_{cm}^2}{(m_Z c^2)^4} . \tag{2.22}$$

If one substitutes $E_{cm}^2 = (1/2)m_e c^2 E_{lab}$ for neutrinos with $E_{lab} = 10\,\text{MeV}$, one finds around $3.5 \times 10^{-18}\,\text{fm}^2$. An exact calculation yields another factor, $1/(96\sin^4\theta_W \cdot \cos^4\theta_W) = 0.30$. One sees that our very simple calculation, in which we corrected for the difference to the electromagnetic interaction only through the $Z^0$ mass, works very well!

The experimental detection of neutrino-electron scattering at low energies $E_\nu \approx 10\,\text{MeV}$ is not easy. The cross-section, $3.5 \times 10^{-18}\,\text{fm}^2$, is small! For comparison: the cross-section for Thomson scattering off a hydrogen atom is $\approx \pi r_e^2 \approx 3.3\,\text{fm}^2$; the typical hadronic cross-section corresponds to the hadronic size, $\approx 1\,\text{fm}^2$.

The strongest available neutrino source is the sun – nuclear reactors, on the other hand, produce antineutrinos! The flux of solar neutrinos is, e.g., measured in a Cherenkov detector filled with 32 000 tonnes of water in Kamiokande (Japan). Neutrinos with an energy of 5.5 MeV deliver a sufficiently large recoil to electrons that their Cherenkov light can be detected.

## 2.4  Neutrino-Quark Scattering

Of course, neutrinos also scatter off quarks by $Z^0$ exchange. Scattering without charge exchange can only be measured by detecting the "jet" originating from the recoil quark. It is experimentally easier to analyse elastic scattering with charge exchange, which is transmitted by $W^\pm$ bosons. To describe neutrino scattering off quarks, we can directly translate the graphs of Fig. 2.1 if we replace the electron e by a neutrino but keep e'. The two quarks, q and q', in the scattering $\nu + q \rightarrow \ell^- + q'$ have different flavours. In 1987, the famous supernova SN1987A was observed in the Large Magellanic cloud. In the Kamiokande detector, 11 antineutrinos were observed that had been generated in this stellar explosion. Where did these antineutrinos come from? Primarily, neutrinos are produced in the collapse of the iron core of the supernova, namely through the reaction $p + e^- \rightarrow n + \nu_e$.

Through the collapse, though, the core is heated up and radiates thermal $\nu\bar{\nu}$ pairs with energies of 3–5 MeV. These thermal antineutrinos can be registered in a detector via $\bar{\nu}_e + p \rightarrow e^+ + n$ (Fig. 2.5). Let us estimate the order of magnitude of the cross-section for this reaction as liberally as with (2.22)

$$\sigma(\bar{\nu}_e p \rightarrow e^+ n) \approx 4\pi \frac{4(\alpha\hbar c)^2 E_{cm}^2}{(m_W c^2)^4} \approx 3 \times 10^{-16}\,\text{fm}^2 \tag{2.23}$$

An exact calculation contributes a further factor, $1/(8\sin^4\theta_W) \approx 2$ to this result.

**Fig. 2.5** Antineutrino
scattering off a proton with
charge transfer

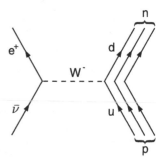

The cross-section for charge exchange is two orders of magnitude larger than
that for neutrino-electron scattering (2.22). The difference lies in $E_{cm}$, which is
significantly higher for scattering off a heavy nucleon.

### 2.4.1  Weak Potential

The scattering amplitude in the Born approximation can be viewed as the Fourier
transformation of the potential. On the other hand, we can extract the weak potential
from the scattering amplitude. The corresponding potential has the Yukawa form,

$$V_W = \frac{\alpha_W \hbar c}{r} e^{(-m_W c/\hbar)r} . \tag{2.24}$$

At low energies, the scattering amplitude is the volume integral of the potential,

$$\int V_W(r) d^3 r = \frac{4\pi \alpha_W (\hbar c)^3}{(m_W c^2)^2} = 4\sqrt{2} G_F . \tag{2.25}$$

The Fermi constant $G_F = 90\,\text{eV}\,\text{fm}^3$ can thus be viewed as the volume integral of
the weak potential. The factor $4\sqrt{2}$ arises from the historical normalisation of $G_F$.

## Literature

I.J.R. Aitchinson, A.J.G. Hey, *Gauge Theories in Particle Physics* (Hilger, Bristol, 1989)
B. Povh et al., *Particles and Nuclei* (Springer, Berlin, 2004)

# Chapter 3
# Quasi-elastic Scattering – Virtual Photons and Gluons

*In jeden Quark begräbt er seine Nase.*

Mephistopheles in Goethe's Faust

For electron energies $E > 15\,\text{GeV}$ and momentum transfers $Q^2 > 1\,\text{GeV}^2$, the scattering takes place off the constituents of the nucleon. Historically, these constituents were christened partons. The name parton includes all the nucleon constituents observed in high-energy scattering: valence quarks, sea quarks and gluons. The term valence quarks refers to the three quarks that contribute to the baryon number and charge of the nucleon, while the sea quarks are produced in equilibrium with the gluons through pair creation and annihilation. This separation is, though, somewhat artificial. In lepton scattering, one only sees the quarks, which have both an electric and a weak charge.

In a proton, the quarks are bound and move inside the limit set by confinement with Fermi momenta corresponding to this limit. A large momentum transfer guarantees that scattering takes place in such a short time that it is safe to neglect interactions between the quarks during the collision. Thus, to a good approximation, we can consider the scattering as being off a free, though not stationary, quark. Historically, this quasi-elastic scattering regime was called deep inelastic scattering. We, however, prefer to call such scattering quasi-elastic scattering.

The masses of the bare light quarks are of the order of magnitude of $10\,\text{MeV}/c^2$. Confined in a volume with diameter $\approx 1$ fm, the quarks with such small masses must be understood as relativistic particles. Statistical descriptions are appropriate for high-energy processes of relativistic many-body systems: in the case of the nucleon, we will describe the partonic structure in terms of the momentum distribution of the quarks and gluons (without referring to a wave function).

At large momentum transfer, we may apply the perturbative field theory of the strong interaction (QCD). Quasi-elastic scattering needs to be formulated in a Lorentz

© Springer-Verlag GmbH Germany 2017
B. Povh and M. Rosina, *Scattering and Structures*,
Graduate Texts in Physics, DOI 10.1007/978-3-662-54515-7_3

invariant fashion. We also need, though, a clear interpretation of the formal theory. This can be taken from the interpretation of QED provided by the Weizsäcker–Williams method's picture of virtual photons. We believe that a brief summary of this method, which in the case of electrodynamics, is conceptually very simple, is extremely useful for an understanding of the partonic description. Furthermore, one can very nicely bring out the difference between QED and QCD by comparing the photon field of an electric charge with the gluon field of a strong charge.

This chapter is a bit longer than the others of this book because, as far as we know, a similar introduction to the strong interaction was not available in textbooks.

## 3.1   Virtual Weizsäcker–Williams Photons

Bremsstrahlung is usually understood as the radiation that accompanies the braking of an electron in the Coulomb field of an atomic nucleus. But both bremsstrahlung and other processes can also be considered in a frame of reference in which the electron is at rest. This is called the virtual quantum method. We will see that this alternative is very well suited to strongly interacting systems.

In the electron's rest frame, the proton approaches the electron with a large energy, $E \gg Mc^2$. The Coulomb field of a moving charge $+e$ (for a proton) with mass $M$ and energy $E$ is Lorentz contracted, as is symbolically shown in Fig. 3.1. The transverse electric field is increased through Lorentz contraction by a factor $\gamma = E/Mc^2$. At a distance $b$ transverse to the direction of motion, we have

$$E_\perp = \frac{e\gamma}{4\pi\varepsilon_0 b^2} \, . \tag{3.1}$$

Observers at a point P (Fig. 3.1) see the charge pass them by as an electric and magnetic pulse. Below, we will only consider the transverse component of the electric pulse because it alone is important for our needs. The duration of the pulse is

$$\Delta t \approx \frac{b}{\gamma c} \, , \tag{3.2}$$

**Fig. 3.1** The spherically symmetric Coulomb field of a charge at rest is Lorentz contracted when it moves. The transverse electric field is amplified by a factor $\gamma = E/Mc^2$

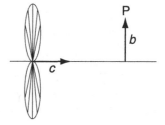

**Fig. 3.2** Simple estimation of the pulse (*rectangular shape $E_\perp \times \Delta t$, thin line*) and its realistic shape (*thick line*)

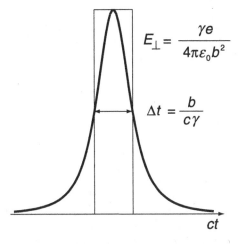

$$E_\perp = \frac{\gamma e}{4\pi\varepsilon_0 b^2}$$

$$\Delta t = \frac{b}{c\gamma}$$

$ct$

**Fig. 3.3** Distribution of the energy flux against $\omega$; the *thin line* for a sharp cut-off frequency and the *thick line* for a more realistic behaviour

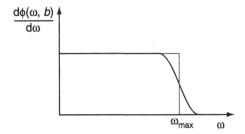

$\dfrac{d\phi(\omega, b)}{d\omega}$

$\omega_{\max}$        $\omega$

where we may always assume that the speed of the charge is $c$ and, on transforming into the laboratory frame, the time scale picks up a factor of $\gamma$ in the denominator. The form of the electric pulse is shown in Fig. 3.2.

We can straightforwardly estimate the dependence of the energy flux on the frequency without exactly calculating the Fourier transform. The energy pulse is – as with $E_\perp$ – of a very short duration, and its energy flux is

$$\Phi = c\varepsilon_0 \int_{-\infty}^{+\infty} E_\perp^2 \, dt \,. \tag{3.3}$$

The Fourier transform of a delta-like pulse ($\Delta t \to 0$) is constant. For a finite width, $\Delta t$, the spectrum is cut off at a maximal frequency, $\omega_{\max} = 1/\Delta t = \gamma c/b$ (see Fig. 3.3).

To compare the virtual photon spectrum with the distribution of photons in the electron (the em structure function), we have to quantise the energy flux and we introduce the usual variable of QCD, $Q \propto \hbar/b$:

$$\frac{d\Phi(\omega, b)}{d\omega} d\omega db^2 \propto \hbar\omega \Gamma(\hbar\omega, Q^2) d(\hbar\omega) dQ^2 \,. \tag{3.4}$$

**Fig. 3.4** The function $\Gamma(x, Q^2)$ yields the number of bremsstrahlung photons in an interval $d(\hbar\omega)$ at fixed $Q^2$

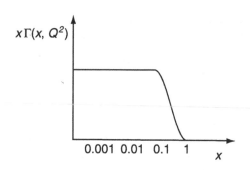

In order to show the analogy with the parton picture (next Section) we use the notation $x = \omega/\omega_{\text{max}}$ where $x$ means here the fraction of the energy carried by the electromagnetic field while in the parton picture (3.6) $x$ means the fraction of the four-momentum carried by quarks or gluons. We will directly compare the distribution of the virtual photons $x\Gamma(x, Q^2)$ (Fig. 3.4) with the gluon structure function $xG(x, Q^2)$.

The bremsstrahlung spectrum at fixed $Q^2$ is obtained by multiplying the function $\Gamma(x, Q^2)$ by the Compton cross-section. Experimentally, the spectrum is determined via the detection of a coincident bremsstrahlung photon and the recoil electron. From Fig. 3.4, one sees that the *form* of the "structure function" of the virtual photons is independent of $Q^2$, i.e., $x\Gamma(x, Q^2 = const) = Const$. The Const. increases with $Q^2$, and so does the number of photons.

Let us explicitly write out the cross-section for soft X-rays. In this case, we may approximate the electron-photon cross-section by the Thomson formula (1.16)

$$\frac{d\sigma(\omega, \theta)}{d\omega d\Omega} \propto Z^2 r_e^2 \frac{1 + \cos^2\theta}{2} \int db^2 \frac{\Phi(\omega, b)}{d\omega} . \tag{3.5}$$

The bremsstrahlung spectrum is the integral over all possible values of $b^2$ (or all momentum transfer $Q^2$).

In the next section, we will see that (3.5) may be directly carried over to quasi-elastic electron-quark scattering.

## 3.2  Virtual Bjorken–Feynman Partons – Deep Inelastic Scattering

Let us consider a proton in a rapidly moving system. We neglect transverse momenta, as we previously did with the longitudinal field components in the electromagnetic case. The total energy of the proton is carried by partons. Each parton bears a fraction $x$ of the total energy, total momentum and mass (see Table 3.1 and Fig. 3.5).

Let us initially discuss the relation between quasi-elastic scattering off a nucleon with the virtual photon picture. Gluons pose a particular difficulty because the static

**Table 3.1** Proton and parton kinematical quantities in a rapidly moving system: $p_L$ denotes the longitudinal momentum component and $p_T$ the transverse components

|          | Proton | Parton |
|----------|--------|--------|
| Energy   | $E$    | $xE$   |
| Momentum | $p_L$  | $xp_L$ |
|          | $p_T = 0$ | $p_T = 0$ |
| Mass     | $M$    | $m = (x^2 E^2 - x^2 p_L^2)^{1/2} = xM$ |

**Fig. 3.5** Partons

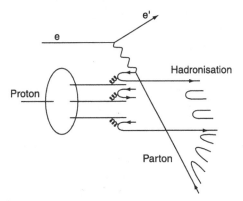

gluon field cannot be represented analytically because of the size of the strong coupling constant $\alpha_s$ and/or confinement. We assume, though, that, in a rapidly moving system, the gluon field may be viewed as virtual gluon quanta. A further complication is that gluons have neither an electric nor a weak charge. One cannot directly observe them in lepton scattering. The partons with an electric and weak charge are the quarks. Thus, we will first consider scattering off quarks.

## 3.2.1   Electron Scattering Off Quarks

The $x$ variable can be expressed in terms of Lorentz invariant quantities. The condition for elastic scattering off partons with mass $m = xM$ follows from (2.9),

$$x = \frac{Q^2}{2M\nu}. \tag{3.6}$$

To get the prefactors right in the definition of the structure function, we have to introduce the $x$ variables in a somewhat more formal way.

The probability of finding a parton with fraction $x$ of the total momentum of the proton is given by the distribution $q_i(x)$. The index $i$ denotes the flavour of the parton and thus its charge. The charge of the quarks is expressed in units of the elementary charge, $z_i e$. The cross-section for quasi-elastic scattering may then be described as an incoherent sum of elastic scatterings off the quarks (2.12),

$$
\left( \frac{d\sigma(eq \to eq)}{dE' d\Omega} \right)_{nucleon}
$$
$$
= \frac{d\sigma_{Mott}}{d\Omega} \sum_i z_i^2 q_i(x) \cdot \left( 1 + 2\frac{Q^2}{4m_i^2 c^2} \tan^2 \frac{\theta}{2} \right) \cdot \delta \left( \nu - \frac{Q^2}{2m_i} \right) . \tag{3.7}
$$

Introducing a new variable $\xi = Q^2/2M\nu$ and substituting it into the delta function yields

$$
\delta \left( \nu - \frac{Q^2}{2m} \right) = \delta \left[ \frac{\nu}{x}(x - \xi) \right] = \frac{x}{\nu} \delta(x - \xi) . \tag{3.8}
$$

The only contribution to the cross-section comes from quarks which carry the fraction $x$ of the total momentum is

$$
x = \xi = \frac{Q^2}{2M\nu} . \tag{3.9}
$$

The final expression is then

$$
\left( \frac{d\sigma(eq \to eq)}{dE' d\Omega} \right)_{nucleon}
$$
$$
= \frac{d\sigma_{Mott}}{d\Omega} \left( \frac{\sum_i z_i^2 x q_i(x)}{\nu} + \frac{\sum_i z_i^2 q_i(x)}{M} \tan^2 \frac{\theta}{2} \right) . \tag{3.10}
$$

It is usual to denote the incoherent sums over the contributions from the individual quarks to the cross-section as structure functions. The structure function that determines the spin-flip part of the cross-section for quasi-elastic scattering is normalised as

$$
F_1 = \frac{1}{2} \sum_i z_i^2 q_i(x), \tag{3.11}
$$

and that which describes the Coulombic part is – just as for spinless quarks

$$
F_2 = \sum_i z_i^2 x q_i(x) . \tag{3.12}
$$

The reader should not confuse the traditional notation for structure functions $F_1$ and $F_2$ with the form factors $F$ which are written without an index (Chap. 2).

The interpretation of the structure functions is, as previously mentioned, especially clear in a frame in which the proton is moving rapidly. In such a frame, the function $2F_1$ gives the probability of finding a parton with fraction $x$ of the total momentum of the proton. The function $F_2$ is the same probability multiplied by $x$. The analogy between (3.10) and (3.5) is evident. The electron-quark cross-section corresponds to the photon-electron cross-section and the quark structure function to the photon structure function.

It may seem surprising to introduce the strong interaction with the virtual photons method of Weizsäcker–Williams. The method was of interest in the time when the photon-electron interaction was treated semiclassically and it was suitable for quantising a classical system. Later, it was replaced by the development of the field theory of QED. In QCD too, "proper" theorists calculate, as far as they can, in a Lorentz invariant, field-theoretic formalisms. There are two reasons why we have chosen the Weizsäcker–Williams method to study the strong interaction. On the one hand, this method is very straightforward because the gluons that are measured in experiments can be interpreted as bremsstrahlung gluons. On the other hand, one of the most important theoretical methods that can be applied to the nonperturbative QCD domain, the light-cone method, is not much more than a somewhat formalised Weizsäcker–Williams method.

### 3.2.2  Neutrino Scattering Off Quarks

Measurements of the structure functions in quasi-elastic neutrino scattering are interesting because the cross-sections of neutrinos and antineutrinos off quarks and antiquarks differ. Experimentally, the reactions have been most thoroughly investigated in reactions with muon neutrinos and antineutrinos,

$$\nu_\mu + q_{d,s,\bar{u}} \rightarrow \mu^- + q_{u,c,\bar{d}} \tag{3.13}$$

and

$$\bar{\nu}_\mu + q_{u,\bar{d},\bar{s}} \rightarrow \mu^+ + q_{d,\bar{u},\bar{c}} . \tag{3.14}$$

This is because pure, high-energy beams are only available for muon neutrinos. They are produced in so-called tertiary beams after pion decays, $\pi^+ \rightarrow \nu_\mu + \mu^+$ and $\pi^- \rightarrow \bar{\nu}_\mu + \mu^-$. At CERN, pions are produced from 400 GeV protons. Pions and kaons are kept bunched for a distance of about 300 m and are braked on a graphite target. The neutrinos from the decay are kinematically directed forward. The spectrum is wideband, with a maximum at 26 GeV and a higher-energy tail up to around 150 GeV. The quasi-elastic scattering is identified by the measurement of the energy of the produced particles in a calorimeter. From the muon's kinematics and the energy of the hadrons produced in the scattering, one can determine both the momentum and energy transfer. Because of the conservation of fermion helicity at at high energies,

**Fig. 3.6** Scattering of
neutrinos off quarks takes
place in the state $S_z = 0$. The
angular distribution is
isotropic. Off antiquarks
($S_z = -1$), the angular
distribution is proportional to
$\cos^2 \theta$

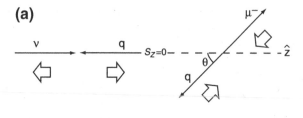

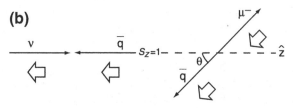

there is a difference between the neutrino and antineutrino cross-sections off quarks.
We can estimate this difference assuming that quarks are massless. In the centre of
mass system, the neutrinos and quarks have a total spin component $S_z = 0$. This is
because, due to the parity violation of the weak interaction, both the neutrino and
the quark have negative helicity.

Because, in neutrino scattering at these energies, only s-wave scattering takes
place, the final state must also have $S_z = 0$. This is the case for the scattering of neutrinos off quarks and of antineutrinos off antiquarks for all scattering angles (Fig. 3.6a).
This is not the case for the scattering of neutrinos off antiquarks (Fig. 3.6b). The
spin component before scattering is $S_z = -1$ (for antineutrinos off quarks, it is
$S_z = +1$). The scattering amplitude depends on the scattering angle $\theta$ and is proportional to $\cos \theta$, so the cross-section is then proportional to $\cos^2 \theta$. Because the
average value of $\langle \cos^2 \theta \rangle = 1/3$, one expects the ratio between both cross-sections
to be 3:1. This ratio depends especially on the kinematics but also on $x$ (see (3.6)). An
exact calculation confirms that the ratio of the neutrino-quark and antineutrino-quark
cross-sections averaged over $x$ is indeed roughly 3:1.

From a comparison of quasi-elastic scattering of neutrinos and antineutrinos, one
can determine the prevalence of quarks and antiquarks in the nucleon. In Fig. 3.7, the
distributions of valence and sea quarks for $Q^2 \approx 5\,\mathrm{GeV}^2/c^2$ and $Q^2 \approx 50\,\mathrm{GeV}^2/c^2$
are shown.

The area below the valence quark distribution measures the valence quarks
weighted by the square of their charges and their fraction $x$ of the total momentum,

$$\int_0^1 \frac{F_2(x)}{x}\,\mathrm{d}x \approx \int_0^1 \left[ \left(\frac{2}{3}\right)^2 + \left(\frac{2}{3}\right)^2 + \left(\frac{1}{3}\right)^2 \right] q(x)\mathrm{d}x = 1 . \qquad (3.15)$$

We have here assumed that the u and d quarks have the same distribution.

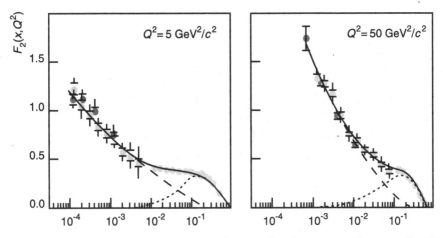

**Fig. 3.7**  The structure function $F_2(x)$ for (*left*) $Q^2 \approx 5 \mathrm{GeV}^2/c^{-2}$ and (*right*) $Q^2 \approx 50 \mathrm{GeV}^2/c^{-2}$. The separation between valence quarks (*dots*) and sea quarks (*dashes*) was determined from neutrino scattering

### 3.2.3   Gluon Bremsstrahlung

As mentioned above, one only sees in quasi-elastic lepton scattering those partons that have either an electric or a weak charge. The presence of the gluons can only be indirectly determined. The sum of the quark momenta in quasi-elastic scattering is only about half the total momentum of the nucleon and the missing half is ascribed to the gluons.

Bremsstrahlung gluons are always observed when a strong charge is accelerated. They are manifested as hadronic jets. In $e^+e^-$ annihilation into a quark antiquark pair, for example, one observes two opposed hadronic jets, which correspond to the hadronisation of the quarks. Sometimes a "bremsstrahlung gluon" makes itself manifest in the process as an additional third jet (Fig. 3.8).

Gluon bremsstrahlung has been most thoroughly investigated in quasi-elastic scattering, as we would like to briefly describe. The value of $Q^2$ also defines the spatial resolution, $\Delta r$, with which one investigates the structure of the object.

$$\Delta r \propto \frac{\hbar c}{Q}. \tag{3.16}$$

The structure function evidently depends on $Q^2$, the resolution of the measurement, as can be seen from Fig. 3.7. This $Q^2$ dependence is brought out in Fig. 3.9. For a poor resolution, one measures the momenta of the partons inside the volume defined by the resolution. The better the resolution, the more partons are measured. If one knows the coupling constant, one can calculate the probabilities for the processes that lead to the splitting of quarks and gluons. The "*splitting function*" probabilities are graphically represented in Fig. 3.10.

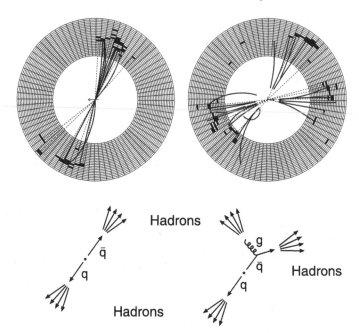

**Fig. 3.8** Typical two- and three-jet events measured at the JADE detector in the PETRA $e^+e^-$ storage ring

**Fig. 3.9** The interaction of a photon with a quark, which radiates a gluon. For the smaller $Q_1^2$, the quark and gluon are not separated. For the larger $Q_2^2$, the resolution is increased and one measures the momentum fraction of the quark without the gluon. The logarithmic dependence of the resolution follows from (3.17)

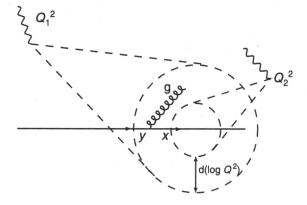

This system of coupled equations describes the $Q^2$ dependence of the structure functions very well. In the measurement, one can only determine the structure functions of the quarks. The gluons have neither an electric nor a weak charge. However, from the $Q^2$ dependence of the quark structure functions, $F_2(x, Q^2)$, the gluon structure, $G(x, Q^2)$, can be determined with the help of the equations in Fig. 3.10. For all $Q^2$ values, the sum of the quark and gluon momenta must be equal to the total momentum of the nucleon. From this condition, the gluonic structure functions may be determined.

$$\frac{d}{d(\ln Q^2)} \left( \begin{array}{c} F_2^N(x,Q^2) \\ xG(x,Q^2) \end{array} \right)$$

$$= \frac{\alpha_s(Q^2)}{2\pi} \left( \begin{array}{cc} P_{qq} & P_{qg} \\ P_{gq} & P_{gg} \end{array} \right) \left( \begin{array}{c} F_2^N(y,Q^2) \\ yG(y,Q^2) \end{array} \right)$$

where

$$\left( \begin{array}{cc} P_{qq} & P_{qg} \\ P_{gq} & P_{gg} \end{array} \right) = \left( \begin{array}{cc} & \\ & \end{array} \right)$$

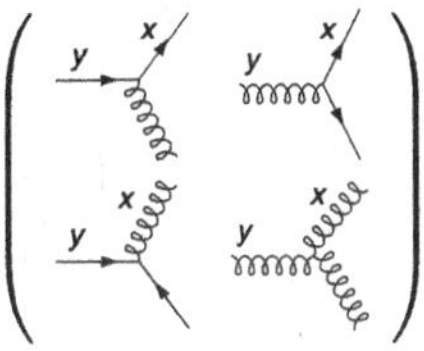

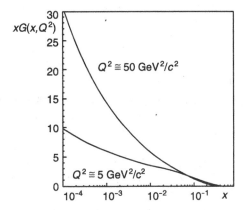

**Fig. 3.10**   The function $P_{qq}$ is the probability that the quark radiates a gluon when one improves the resolution of the measurement by $d(\ln Q^2)$; $P_{qg}$ the probability that a quark with $x$ is pair produced; $P_{gq}$ the probability that a gluon with $x$ is created in quark annihilation; $P_{gg}$ the probability of gluon fission

**Fig. 3.11**   The gluon structure function for $Q^2 = 5\,\mathrm{GeV^2}/c^2$ and $Q^2 = 50\,\mathrm{GeV^2}/c^2$. The greater the value of $Q^2$, the more soft gluons there are

In Fig. 3.11, the gluon structure functions are shown for $Q^2 = 5\,\mathrm{GeV^2}/c^2$ and $Q^2 = 50\,\mathrm{GeV^2}/c^2$.

These gluonic structure functions should be compared with the photonic $x\Gamma(x, Q^2)$ (cf. Fig. 3.4). Both processes, the bremsstrahlung of photons and of gluons, are subject to the same general laws. If gluons themselves did not carry a strong charge, both bremsstrahlung spectra would appear identical. The difference comes from the self-coupling of the gluons. Thus, High-energy gluons split into more lower energy ones and the gluonic spectrum is shifted to lower $x$. The better the resolution, the more the total momentum is carried by soft gluons.

## 3.3  Coupling Constants

Quantum chromodynamics (QCD) is the generally accepted theory of the strong interaction. Quarks carry the strong charge, colour. The interaction between the quarks is transmitted by gluons. Gluons themselves carry the strong charge and couple to each other. The strength of the coupling of quarks to gluons and of gluons to gluons is given by the "coupling constant" $\alpha_s$. It is, however, strongly dependent on $Q^2$. At $Q^2 \approx 10^4 \text{GeV}^2 c^{-2}$, it has a value of around 0.12, at $Q^2 \approx 10^2 \text{ GeV}^2 c^{-2}$, about 0.16 and for $Q^2 \approx 1 \text{ GeV}^2 c^{-2}$, it is about 0.5. The $Q^2$ dependence is logarithmic and given by the following expression:

$$\alpha_s(Q^2) = \frac{12\pi}{(33 - 2n_f) \cdot \ln(Q^2/\Lambda^2)} . \tag{3.17}$$

Here, $n_f$ denotes the number of quark flavours involved. Because virtual quark-antiquark pairs of heavy quarks have only a very short Lifetime, their separation from a struck quark is so small that they can first be resolved for very large values of $Q^2$. For $Q^2$ in the region of $1\, GeV^2 c^{-2}$, one expects $n_f \approx 3$ and $n_f = 6$ as $Q^2 \to \infty$. The only free parameter in QCD, $\Lambda$, must be experimentally determined; its value is around $250 \text{ GeV}/c$. For comparison: QED also has one free parameter, the fine-structure constant $\alpha$, which is experimentally very easily extracted from the Thomson cross-section.

The experimental verification of QCD cannot be as elegantly performed as is possible in the case of QED. Also, the precision with which QED is tested will never be achieved in QCD. The quarks and gluons are confined both before and after scattering. One can only observe hadronised quarks and gluons after the interaction. Because the hadrons are bunched (jets) in the direction of the scattered quarks and gluons, one can well reconstruct the elementary process.

Let us try to make out the origin of the expression (3.17). The $Q^2$ dependence of the coupling constant is indeed not only a property of the strong interaction but rather a general property of all interactions and a consequence of vacuum polarisation.

### 3.3.1  Electromagnetic Coupling Constant $\alpha$

The attraction between positive and negative charges is not exactly given by the Coulomb law. At short separations, $r < \lambda_e$, the effective charge increases because the charge polarises the virtual electron–positron pairs. This polarisation is parallel to the electric field. The polarisation vector points in the direction $\mathbf{r}$ when the charge is positive. This must be the case, as the virtual electron–positron pairs (Fig. 3.12) distribute themselves such that positive charge is forced out away from the centre.

We can roughly estimate the size of this correction. The value of the loop in Fig. 3.12 is well known to depend on $\log Q^2$. The integration is from 0 to $\infty$, and the

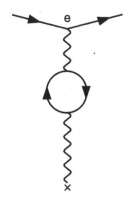

**Fig. 3.12** The lowest correction to the Coulomb law. The electron–positron pairs distribute themselves such that identical charges repel each other

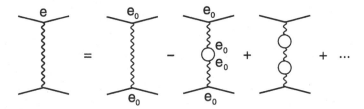

**Fig. 3.13** Higher order corrections to the vacuum polarisation form a geometric series

**Fig. 3.14** Graphical depiction of the sum of the loops that contribute to the vacuum polarisation

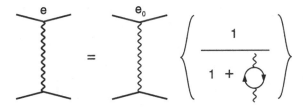

value of the integral diverges. We are, though, only interested in the $Q^2$ dependence of the coupling constant, when its value is already known from a measurement at a momentum transfer $\mu^2$. Then the value of the loop is (Fig. 3.12)

$$-\frac{\alpha}{3\pi}\ln\left(\frac{Q^2}{\mu^2}\right). \tag{3.18}$$

The final result for the coupling constant is found when we also include the higher order corrections of Fig. 3.13.

The sum of the corrections shown in Fig. 3.13 is graphically represented in Fig. 3.14 and can be analytically written as the sum of the geometrical series of powers of (3.18),

$$\alpha(Q^2) = \frac{\alpha(\mu^2)}{1 - [\alpha(\mu^2)/3\pi]\ln(Q^2/\mu^2)}. \tag{3.19}$$

This formula holds for $Q, \mu \gg m_e c$, and the value of $\alpha$ is given at a suitable scale. At separations $r \geq \lambda_e$, the vacuum polarisation is negligible. Values for $\alpha$ have been measured up to $Q^2 \approx 10^4 \, GeV^2 c_0^{-2}$, and they are in agreement with the expression (3.19).

### 3.3.2  Strong Coupling Constant $\alpha_s$

The vacuum polarisation for strongly interacting particles is treated in just the same way as with the particles of the electromagnetic interaction. The only difference is that not simply quark–antiquark pairs (Fig. 3.15) contribute to the polarisation, but it also involves gluonic loops. The contribution to the polarisation from quark–antiquark pairs screens the strong charge, just as electron–positron pairs do for electric charges (Fig. 3.15: $g \rightarrow q\bar{q}$). This contribution is weighted by the number of flavours, $n_f$. The self-coupling of the transverse gluons (Fig. 3.15: $g \rightarrow g_T g_T$) yields the same polarisation. However, one has shown from QCD that the dominant term (Fig. 3.15: $g \rightarrow g_C g_T$), the self-coupling of gluons to transverse and Coulombic gluons, forces the strong charge outward.

Analogously to the sum of the geometric series (Fig. 3.13), the expression for $\alpha_s(Q^2)$, compared to a reference value at $Q^2 = \mu^2$, is

$$\alpha_s(Q^2) = \frac{\alpha_s(\mu^2)}{1 + [\alpha_s(\mu^2)12\pi](33 - 2n_f)\ln(Q^2/\mu^2)}. \tag{3.20}$$

The equation (3.17) is obtained by substituting the commonly used scale $Q^2$ into (3.20)

$$\Lambda^2 = \mu^2 \exp\left[\frac{-12\pi}{(33 - 2n_f)}\alpha_s(\mu^2)\right]. \tag{3.21}$$

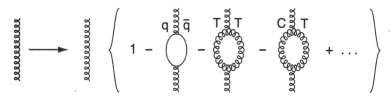

**Fig. 3.15**  Contributions from gluonic loops to the vacuum polarisation: gluon $g \rightarrow q\bar{q}$, $g \rightarrow g_T g_T$, $g \rightarrow g_C g_T$

### 3.3.3  *Weak Coupling Constant* $\alpha_W$

The weak bosons, $W^{\pm,0}$ (see Chap. 16), carry weak isospin, the effect of self-coupling dominates the vacuum polarisation and the strength of $\alpha_W$ decreases for larger $Q^2$.

## Literature

F. Halzen, A.D. Martin, *Quarks and Leptons* (Wiley, New York, 1984)

J.D. Jackson, *Classical Electrodynamics* (Wiley, New York, 1975)

B. Povh et al., *Particles and Nuclei* (Springer, Berlin, 2015)

# Chapter 4
# The Hydrogen Atom – The Playground of Quantum Mechanics

> *Das Atom der modernen Physik kann allein durch eine partielle Differentialgleichung in einem abstrakten vieldimensionalen Raum dargestellt werden. Alle seine Eigenschaften sind gefolgert; keine materiellen Eigenschaften können ihm in direkter Weise zugeschrieben werden. Das heißt jedes Bild des Atoms, das unsere Einbildung zu erfinden vermag, ist aus diesem Grunde mangelhaft. Ein Verständnis der atomaren Welt in jener ursprünglichen sinnlichen Weise ist unmöglich.*
>
> Heisenberg in 1945

The hydrogen atom is the simplest atomic system. It may be described as a one-body system to a very good accuracy and can be solved analytically in this way. It is thus suitable for testing quantum mechanics. Furthermore, precision tests on the hydrogen atom still provide the most accurate tests of quantum electrodynamics (QED).

All the properties of the hydrogen atom are determined by the charge $e$ of the electron, its mass, $m_e$, and Planck's constant, $\hbar$. We will use the dimensionless Fine-structure constant $\alpha = e^2/(4\pi\varepsilon_0\hbar c)$ as the coupling constant of the electromagnetic interaction.

## 4.1 Level Diagram

### 4.1.1 Semiclassical

The electron moves in the Coulomb field of the proton with an average potential energy $\bar{V} = -\alpha\hbar c/\bar{r}$. Here, $\bar{r}$ is the radius of the classical orbit of the electron around the proton (actually, $\bar{r} = \langle 1/r \rangle^{-1}$).

© Springer-Verlag GmbH Germany 2017
B. Povh and M. Rosina, *Scattering and Structures*,
Graduate Texts in Physics, DOI 10.1007/978-3-662-54515-7_4

The average kinetic energy of the electron is $\bar{K} = \bar{p}^2/2m_e$, and $\bar{p}$ is its average momentum (more exactly, $\sqrt{\langle p^2 \rangle}$). The smearing of the position and momentum in the ground state of the atom must obey the uncertainty relation. The uncertainty relation is an inequality. If one uses it as an equality ($\Delta r \cdot \Delta p = k\hbar$), the factor $k$ in front of $\hbar$ depends on the potential. To get quantitative results for the case of the Coulomb potential, one must require $\bar{r}\bar{p} = \hbar$. This is reminiscent of the de Broglie rule, which states that the circumference of stable orbits must be an integer multiple of the de Broglie wavelength, $\lambda = h/p$.

A short clarification is in order: in the bound state, wavelength is not well defined, but in a quantum state of size $\bar{r}$, without any nodes, $\bar{r} \approx \lambda$. To show this, one must perform a Fourier analysis of the Schrödinger wave function. It emerges that, in objects of size $\bar{r}$, the main contributions come from waves with $\lambda = \bar{r}$. As we will see later in the case of the hydrogen atom, $\bar{r}$ is the radius at which the electron density distribution multiplied by $r^2$ has its maximum. We also call this radius the most probable radius.

With the help of the uncertainty relation, the average kinetic energy may be written as

$$\bar{K} = \frac{\hbar^2}{2m_e\bar{r}^2} \,. \tag{4.1}$$

The ground state radius, $\bar{r}$, is found from the condition that the total energy of the system

$$E = -\frac{\alpha \hbar c}{\bar{r}} + \frac{\hbar^2}{2m_e\bar{r}^2} \tag{4.2}$$

is minimal: $dE/d\bar{r} = 0$. The radius at the minimum is called the Bohr radius, $a_0$:

$$a_0 = \frac{\hbar c}{\alpha m_e c^2} = \frac{\lambda_e}{\alpha} \,. \tag{4.3}$$

Here, $\lambda_e$ is the Compton wavelength of the electron. The binding energy of the hydrogen atom, which is also called the Rydberg constant, Ry, is

$$E_1 = -\frac{\alpha^2 m_e c^2}{2} = -\frac{\alpha \hbar c}{2a_0} := -1\,\mathrm{Ry} = -13.6\,\mathrm{eV}. \tag{4.4}$$

In the ground state, with principal quantum number $n = 1$, the wave function has no nodes and the de Broglie wavelength is $\lambda_1 \approx \bar{r}$. In the first excited state, $n = 2$, the wave function has one node and the wavelength is $\lambda_2 = \bar{r}/2$. For the $n$th state, we have $\lambda_n = \bar{r}/n$. This yields the radii and the binding energies,

$$\bar{r}_n = n^2 a_0 \,; \quad E_n = -\mathrm{Ry}/n^2. \tag{4.5}$$

In Fig. 4.1, the nodes of the wavefunctions are illustrated as standing waves and the familiar picture of the hydrogen level diagram is sketched.

**Fig. 4.1** The hydrogen level diagram in semiclassical approximation; the electron continues to be interpreted as a standing wave

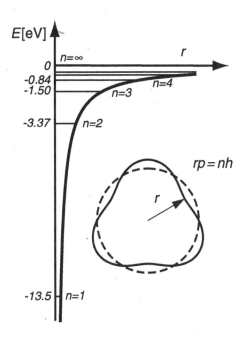

## 4.1.2 Dirac Level Diagram

An excellent understanding of the hydrogen atom is available with the help of the solution of the Dirac equation. This indeed describes the hydrogen atom almost perfectly because it takes the spin and relativistic dynamics of the electron into account. The sole imperfections are to do with the proton's spin, the finite size of the proton and the effects of radiative corrections. We will discuss these below as hyperfine structure and the Lamb shift. The level diagram of the hydrogen atom, calculated from the Dirac equation, is sketched in Fig. 4.2. For comparison, the energy levels found upon ignoring spin and relativity are given. The energy differences are known as relativistic corrections and the spin-orbit interaction. The fine structure splitting $\Delta E_{\mathrm{fs}}$ can be understood as a shift from the nonrelativistic energies; it is (up to $\alpha^2$)

$$\Delta E_{\mathrm{fs}} = -\frac{\alpha^2}{n^3}\left(\frac{1}{j+1/2} - \frac{3}{4n}\right)\mathrm{Ry}. \tag{4.6}$$

It is interesting to note that the states calculated from the Dirac equation depend, apart from the principal quantum number, solely on the total angular momentum, $\mathbf{j} = \boldsymbol{\ell} + \mathbf{s}$. The orbital angular momentum is not a good quantum number.

In the following, we only want to show the plausibility of the orders of magnitude of various shifts from the nonrelativistic energies. This is very educational. One sees how elegantly the Dirac equation encompasses relativistic effects. On the other

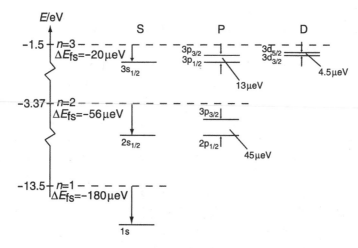

**Fig. 4.2** The level diagram of the hydrogen atom, calculated from the Dirac equation. The energy levels from nonrelativistic calculations are given for comparison

hand, the Dirac equation is only exactly soluble for hydrogen-like atoms and ions. For atoms with more than one electron, there are no exact solutions.

For $n = 1$, one can estimate the relativistic corrections rather accurately in first-order perturbation theory, and we will consider this in detail.

The correction to the kinetic energy is

$$\Delta_K = -\frac{p^4}{8m_e^3 c^2} \,. \tag{4.7}$$

In the ground state, we have

$$E_0 = \frac{p^2}{2m} + V \,, \tag{4.8}$$

and the correction to the kinetic energy may be written as

$$\Delta_K = -\frac{p^4}{8m_e^3 c^2}$$
$$= -\frac{1}{2mc^2}\left[ E_0^2 + 2E_0 \alpha \hbar c \left\langle \frac{1}{r} \right\rangle + \alpha^2 (\hbar c)^2 \left\langle \frac{1}{r^2} \right\rangle \right]. \tag{4.9}$$

Substituting $E_0 = -1\,\text{Ry}$ and taking into account that, in the ground state, $\langle 1/r \rangle = 1/a_0$ and $\langle 1/r^2 \rangle = 2/a_0^2$, one finds

$$\Delta_K = -\frac{5}{4}\alpha^2 \text{Ry}. \tag{4.10}$$

$\Delta_K$ is the contribution from the kinetic energy to the reduction of the ground state energy.

### 4.1.3 Zitterbewegung

The electron cannot be localised with a greater accuracy than its Compton wavelength $\lambda_e$. The positrons, which accompany the electron in the solution of the Dirac equation, annihilate the electron for a moment and create it elsewhere. This leads to a smearing of the electron position, which is historically called "Zitterbewegung".

The stochastic movement of the electron inside a region of size $\lambda_e$ decreases the Coulomb potential at the point $r = 0$. To estimate this correction, we expand the potential in a Taylor series around $\mathbf{r}$,

$$V(\mathbf{r} + \delta\mathbf{r}) = V(\mathbf{r}) + \nabla V \delta\mathbf{r} + \frac{1}{2}\sum_{ij} \nabla_i \nabla_j V \delta r_i \delta r_j + \dots . \qquad (4.11)$$

Because of the vector nature of the smearing, the linear term vanishes on averaging, $\langle \delta\mathbf{r} \rangle = 0$, while the quadratic term has the form

$$\frac{1}{2}\sum_{ij} \nabla_i \nabla_j V \delta r_i \delta r_j \rightarrow \frac{1}{6}\nabla^2 V (\delta\mathbf{r})^2. \qquad (4.12)$$

This term is only nonzero for $\mathbf{r} = \mathbf{0}$ because the Laplace operator applied to the Coulomb potential satisfies the Poisson equation, $\nabla^2(1/r) = -4\pi\delta(\mathbf{r})$, where $\delta(\mathbf{r})$ is, of course, the Dirac delta function.

If we make the approximation $\langle (\delta\mathbf{r})^2 \rangle = \lambda_e^2$, the correction to the Coulomb potential becomes

$$\Delta_D = \frac{1}{6}\lambda_e^2 \alpha \hbar c 4\pi \delta(\mathbf{r}). \qquad (4.13)$$

The energy shift is obtained when we calculate the expectation value of $\Delta_D$. The only contribution comes from $\mathbf{r} = \mathbf{0}$. Therefore, we have to replace the $\delta$-function in (4.13) by the electron probability density at $\mathbf{r} = \mathbf{0}$,

$$|\psi(0)|^2 = \frac{1}{4\pi}\frac{2}{a_0^3} . \qquad (4.14)$$

Our estimate only deviates by 30% from the following correct value, which is known as the Darwin term:

$$\Delta_D = \alpha^2 \text{Ry}. \qquad (4.15)$$

The energy shift in the ground state is then the sum of the two Contributions,

$$\Delta E_{\text{fs}} = \Delta_{\text{K}} + \Delta_{\text{D}} = -\frac{\alpha^2}{4}\text{Ry} = -1.8 \cdot 10^{-4}\,\text{eV}. \qquad (4.16)$$

To obtain precise corrections to excited states, it is necessary to average over the momentum distribution as our liberal estimates yield inexact results.

Consider now the level with $n = 2$. In the $\ell = 0$ state, we have $\Delta_{\text{fs}} = -0.562 \cdot 10^{-4}\,\text{eV}$. For states with $\ell \neq 0$, one must additionally take the spin-orbit coupling into account. This is of a comparable size to the other relativistic corrections and is also proportional to $E_n\alpha^2$.

### 4.1.4   Spin-Orbit Splitting

From (4.6), one can see that states with the same $j$ but different $\ell$ are degenerate. This implies, for the level with $n = 2$, that the relativistic energy shift, $\Delta_{\text{K}} + \Delta_{\text{D}}$, in the $\ell = 0$ state is equal to the sum of both of these and the spin-orbit shift in the ($\ell = 1$, $j = 1/2$) state.

The spin-orbit splitting in the $\ell = 1$ state is

$$\Delta E_{\text{ls}} = \frac{\alpha\hbar}{2\,m_e^2 c}\left\langle\frac{1}{r^3}\right\rangle(\boldsymbol{\ell}\cdot\mathbf{s}). \qquad (4.17)$$

This is easily understood. The magnetic field produced by the proton in the rest frame of the electron is from the Biot–Savart law,

$$\mathbf{B} = \frac{e}{4\pi\varepsilon_0 c^2 r^3}\,\mathbf{r}\times\mathbf{v}. \qquad (4.18)$$

Upon transforming the field into the rotating frame of the atom, the field has to be multiplied by a factor of 1/2 (the Thomas factor). Substituting the angular momentum into (4.18), the field becomes

$$\mathbf{B} = \frac{e}{8\pi\varepsilon_0 m_e c^2 r^3}\,\boldsymbol{\ell}. \qquad (4.19)$$

The $\boldsymbol{\ell}\cdot\mathbf{s}$ shift (4.17) is found by multiplying the magnetic field with the magnetic moment of the electron, $-(e/m)\mathbf{s}$. Because the radial wave function in the state with $n = 2$, $\ell = 1$ is

$$\psi(r) = \frac{1}{\sqrt{24\,a_0^3}}\frac{r}{a_0}\exp\left(-\frac{r}{a_0}\right), \qquad (4.20)$$

the expectation value of $1/r^3$ is found to be

$$\left\langle \frac{1}{r^3} \right\rangle = \frac{1}{24\, a_0^3} \int_0^\infty \exp\left(-\frac{2r}{a_0}\right) \frac{r}{a_0} \frac{dr}{a_0} = \frac{1}{24\, a_0^3}. \tag{4.21}$$

Thus, from

$$\boldsymbol{\ell} \cdot \mathbf{s} = \frac{1}{2}[j(j+1) - \ell(\ell+1) - s(s+1)]\hbar^2$$

$$= \begin{cases} +\dfrac{1}{2}\hbar^2 & \text{for } j = 3/2 \\ -1\hbar^2 & \text{for } j = 1/2 \end{cases} \tag{4.22}$$

the spin-orbit splitting in the $(n = 2, \ell = 1)$ state is

$$\Delta E_{1s}(j = 3/2) - \Delta E_{1s}(j = 1/2) = \frac{1}{4}\alpha^2 E_2 = 0.446 \cdot 10^{-4}\,\text{eV}. \tag{4.23}$$

In Fig. 4.2, the energy shifts are schematically represented for the states $n = 1, n = 2$ and $n = 3$.

## 4.2  Lamb Shift

The Dirac equation perfectly describes fermions insofar as they are elementary particles: it takes properly into account the relativistic effects, the spin and the particle–antiparticle symmetry. But the fine tuning of the energy levels in the atom requires, in addition, considering the radiative corrections—the coupling of the electric charge to the virtual photons and the virtual electron–positron pairs (Fig. 4.3). Surely, also the energy of the free electron undergoes a shift, in fact an infinite one, but is, together with the rest of not known contributions to the electron mass, part of it. Experimentally and theoretically best studied radiative corrections are those on the bound

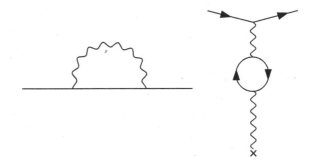

**Fig. 4.3** Electron emits and reabsorbs a virtual photon with a recoil in the meantime. Its potential energy is altered (*left*). The proton charge is screened because of the coupling of the longitudinal photon to the electron–positron pairs (*right*)

electron in the hydrogen atom (Lamb shift) and the magnetic moment of the electron,
i.e., the deviations of the magnetic moment from the Dirac value, $\mu_e = e\hbar/2m_e$. We
discuss only the Lamb shift.

The Lamb shift in the states with $n = 1$, $\ell = 0$ and $n = 2$, $\ell = 0$ is theoretically,
it is believed, known to six figures. In practice, the most accurate data for the Lamb
shift are obtained from measurements of the difference of the 2s and 1s levels in the
hydrogen atom. We will roughly estimate this. The Lamb shift receives primarily
two radiative corrections. The first takes into account the fact that, by emission of a
virtual photon, the electron recoils and smears its position. The second contribution
accounts for the screening of the electric charge by the polarisation of the vacuum
(Fig. 4.3).

Because the first mechanism delivers the major contribution (90%) to the Lamb
shift in the hydrogen atom, we will discuss it alone here. The finite extension of the
proton only generates about 1% of the Lamb shift in hydrogen.

### 4.2.1  Zero-Point Oscillation

There are many ways to treat the virtual photons, preferably by Feyman prescriptions.
But the smearing of the electron position is easier to visualise in the semiclassical
approach of the zero-point oscillation of the electromagnetic field. The recoil of the
electron due to the absorption and emission of the virtual photons is calculated by
taking into account the interaction of the electrons with these zero-point oscillations.

We consider the electromagnetic field to be an incoherent sum of plane waves in
a box of size $L^3$. Each degree of freedom has a zero-point energy of $\frac{1}{2}\hbar\omega$ in the given
phase space, $L^3 4\pi(\hbar\omega/c)^2 \mathrm{d}(\hbar\omega/c)$,

$$\frac{1}{2}\int \mathrm{d}^3x(\varepsilon_0\mathbf{E}^2 + \mu_0^{-1}\mathbf{B}^2) = \frac{1}{2}L^3(\langle\varepsilon_0\mathbf{E}^2\rangle + \langle\mu_0^{-1}\mathbf{B}^2\rangle)$$
$$= 2L^3 \int \frac{4\pi(\hbar\omega/c)^2\mathrm{d}(\hbar\omega/c)}{(2\pi\hbar)^3}\frac{\hbar\omega}{2}. \tag{4.24}$$

The factor 2 in front of the integral comes from the two polarisations of the photon.
Half of the last expression comes from the electric field, from which it follows that

$$\langle\mathbf{E}^2\rangle = \int \frac{\omega^2\hbar\omega}{2\pi^2c^3\varepsilon_0}\,\mathrm{d}\omega. \tag{4.25}$$

This integral is divergent. However, it suffices to estimate the Lamb shift if one
considers the frequency interval between $\hbar\omega_{\min} \approx 2\,\mathrm{Ry}$ and $\hbar\omega_{\max} \approx m_e c^2$. We will
motivate this below.

The field $\mathbf{E}$ accelerates the electron and so smears its coordinates,

$$m_e\delta\ddot{\mathbf{r}} = e\mathbf{E}. \tag{4.26}$$

Let us use this connection to estimate $\langle(\delta\mathbf{r})^2\rangle$. The second time derivative introduces a factor of $(-1/\omega^2)^2$ into the integral over $\omega$ (4.25),

$$
\begin{aligned}
\langle(\delta\mathbf{r})^2\rangle &= \int \left(-\frac{e}{m\omega^2}\right)^2 \frac{\omega^2\,\hbar\omega\,d\omega}{2\pi^2 c^3 \varepsilon_0} = \frac{2\alpha(\hbar c)^2}{\pi m_e^2 c^4} \cdot \int \frac{d\omega}{\omega} \\
&\approx \frac{2\alpha(\hbar c)^2}{\pi m_e^2 c^4} \cdot \ln\frac{\omega_{\max}}{\omega_{\min}}.
\end{aligned}
\tag{4.27}
$$

The coordinate of the electron fluctuates around $\delta\mathbf{r}$, which changes both the kinetic and the potential energy. The alteration of the kinetic energy is the same for both free and bound particles and is included in the renormalisation of the mass. The change in the potential energy is the contribution to the Lamb shift.

Now we have to estimate the relevant ultraviolet and infrared cut-offs for the change in the potential energy.

For the upper bound, we choose the electron mass ($\hbar\omega_{\max} \approx m_e c^2$) because a resolution of the electron position better than the Compton wavelength ($\hbar c/m_e c^2$) is not possible. For the lower bound, we choose a typical atomic energy ($\hbar\omega_{\min} \approx 2\,\mathrm{Ry}$) because bound electrons are not smeared beyond the atomic radius. The ratio of the energies at the upper and the lower cuts is then

$$
\frac{\omega_{\max}}{\omega_{\min}} \approx \frac{m_e c^2}{m_e c^2 \alpha^2} = \frac{1}{\alpha^2}.
\tag{4.28}
$$

The fluctuation of the electron smears the Coulomb potential (see (4.11) and (4.12)),

$$
\Delta_V = \frac{1}{2}\frac{1}{3}\nabla^2 V \langle(\delta\mathbf{r})^2\rangle.
\tag{4.29}
$$

We can apply the Poisson equation and calculate the corrected Coulomb potential,

$$
\nabla^2 V = -4\pi\alpha\hbar c\,\delta(\mathbf{r}),
\tag{4.30}
$$

where $\delta(\mathbf{r})$ is, again, the Dirac delta function. The smearing of the Coulomb potential is then

$$
\langle\Delta_V\rangle = 4\pi\alpha\hbar c\,|\psi_n(0)|^2\langle(\delta\mathbf{r})^2\rangle = 4\pi\alpha\hbar c\,\frac{1}{\pi}\left(\frac{m_e c^2\,\alpha}{\hbar c\,n}\right)^3\langle(\delta\mathbf{r})^2\rangle.
\tag{4.31}
$$

At first order, this corresponds to the shift of the potential energy,

$$
\Delta E_{\mathrm{Lamb}} = \langle\Delta_V\rangle \approx \frac{4}{3\pi}\frac{m_e c^2\,\alpha^5}{n^3}\ln\left(\frac{1}{\alpha^2}\right) = \frac{8}{3\pi}\frac{\mathrm{Ry}\,\alpha^3}{n^3}\ln\left(\frac{1}{\alpha^2}\right).
\tag{4.32}
$$

Our estimate agrees, e.g., for the state $n = 2$ to within 20%.

## 4.3  Hyperfine Structure

Let us continue by considering the interaction between the magnetic moments of the proton and the electron. The magnetic field of a magnetic dipole, e.g., that of a proton, $\mu_p$, is

$$\mathbf{B}(\mathbf{r}) = \frac{\mu_0}{4\pi} \frac{3\mathbf{r}(\mathbf{r} \cdot \boldsymbol{\mu}_p) - r^2 \boldsymbol{\mu}_p}{|\mathbf{r}|^5} + \frac{2\mu_0}{3} \boldsymbol{\mu}_p \delta(\mathbf{r}). \tag{4.33}$$

The dipole–dipole interaction energy can be found by taking the scalar product between the magnetic field of (4.33) and the magnetic dipole moment of the electron and then integrating the electron distribution over all space. The contributions of the first terms thus cancel. Only the contribution of the overlapping moments survives. Only the contact potential, $V_{ss}$, is of significance for the interaction of the magnetic moments of the electron and the proton,

$$V_{ss}(\mathbf{r}) = -\frac{2\mu_0}{3} \boldsymbol{\mu}_p \cdot \boldsymbol{\mu}_e \, \delta(\mathbf{r}). \tag{4.34}$$

From this, the value of the hyperfine splitting is

$$\Delta E_{ss} = -\frac{2\mu_0}{3} \boldsymbol{\mu}_p \cdot \boldsymbol{\mu}_e \, |\psi(0)|^2. \tag{4.35}$$

Only the electrons in the states with $\ell = 0$ have a finite probability of being found at the nucleus. We will only calculate the hyperfine splitting of the 1s in the hydrogen atom. The probability of finding the electron at the position of the proton is from (4.14) $|\psi(0)|^2 = 1/2\pi a_0^3$. The total angular momentum of the atom is denoted by $\mathbf{F}$, and it is the sum of the electron angular momentum and the spin of the nucleus. In the case of the hydrogen atom in the 1s state, we have $\mathbf{F} = \mathbf{s}_e + \mathbf{s}_p$. Because, as is well known,

$$\mathbf{s}_p \cdot \mathbf{s}_e = \frac{1}{2}[F(F+1) - 2s(s+1)]\hbar^2 = \begin{cases} +\dfrac{1}{4}\hbar^2 & \text{for } F = 1 \\[2mm] -\dfrac{3}{4}\hbar^2 & \text{for } F = 0 \end{cases} \tag{4.36}$$

and $\boldsymbol{\mu}_p = 2.973(e/m_p)\mathbf{s}_p$ plus $\boldsymbol{\mu}_e = -(e/m_e)\mathbf{s}_e$, the hyperfine splitting has the value

$$\Delta E_{ss}(F=1) - \Delta E_{ss}(F=0) = \frac{2 \cdot 2.793\mu_0}{3} \frac{e^2(\hbar c)^2}{m_p c^2 m_e c^2} \frac{1}{\pi a_0^3} \tag{4.37}$$

$$= \frac{8\pi \cdot 2.793\alpha^2(\hbar c)^3}{m_p c^2 \cdot m_e c^2} \frac{1}{\pi a_0^3} \tag{4.38}$$

$$= 6 \cdot 10^{-6} \text{ eV}. \tag{4.39}$$

**Fig. 4.4** Complete level diagram of the hydrogen atom including the hyperfine structure splitting

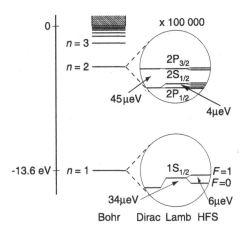

This energy corresponds to the familiar 21-cm radiation, which is emitted from interstellar hydrogen and is easily detected on the earth with antennae. The lifetime of the hyperfine transition is many orders of magnitude too long ($\approx 10^7$ years) to be observed in the laboratory. It is different in the case of interstellar hydrogen: the probability of atomic collisions is sufficiently small there to allow the electromagnetic transition. In Fig. 4.4, the complete level diagram of the H atom, including hyperfine structure splitting, is shown.

## 4.4 Hydrogen-Like Atoms

Negatively charged particles, $\mu^-$, $\pi^-$, $K^-$, $\bar{p}$, $\Sigma^-$, $\Xi^-$, may be successfully placed in the Coulomb field of atomic nuclei. Because the 1s radii behave as $r \propto 1/(mZ)$, a heavy particle will move inside the electron cloud and may very well be viewed as a hydrogen-like atom, though not only with a proton but also with a heavy atomic nucleus at its centre. Atoms with strongly interacting particles bound in the nuclear Coulomb field are well suited for investigations of the particle–nucleus interaction at very low energies. Because the mass of the muon is $\approx 200$ times larger than that of the electron, muons moving inside an atom close to the nucleus are only weakly screened by the electrons.

This is why muonic atoms are suited to measuring the electromagnetic properties of nuclei, as we will now briefly discuss.

### 4.4.1 Muonic Atoms

The binding energies in muonic atoms can be calculated for most states by taking the formulae for the hydrogen atom and replacing both the electron mass by that of the muon and the proton charge by that of the nucleus in question. A significant deviation from the hydrogen formulae estimates is found for the $\ell = 0$ states and, in particular, for the $1s_{1/2}$ ground state. We will demonstrate this for the example of the muonic lead atom. To estimate the ground-state energy of the muonic lead atom, we can take the electric charge in lead as being constant inside the nuclear radius, $R = 7.11\,\text{fm}$.

In a muonic atom with a point nucleus carrying the charge of lead ($Z = 82$), the most probable radius of the muon in the $1s_{1/2}$ state would be

$$a_\mu = \frac{a_0}{Zm_\mu/m_e} = \frac{a_0}{16960} \approx 3.1\,\text{fm}, \qquad (4.40)$$

and its binding energy would be

$$E_1 = -Z^2 \frac{m_\mu}{m_e}\,\text{Ry} \approx -18.92\,\text{MeV}. \qquad (4.41)$$

The experimental result for the binding energy of the $1s_{1/2}$ state in muonic lead is only $E_{1s} = -9.744\,\text{MeV}$.

A muon moving so close to the nucleus feels a strongly modified Coulomb potential because the lead nucleus has a radius of about 7.1 fm, which is comparable with the extension of the muonic wave function. The effective Coulomb potential of a lead nucleus is sketched in Fig. 4.5. We have assumed that the nucleus is a homogenous charged sphere of radius $R$. Inside the nucleus, the potential increases as $r^2/R^3$ and has the form of a harmonic oscillator, while outside the nucleus, a simple $1/r$ dependence holds. At the edge of the nucleus $R$, both functions must have the same value and the same derivative. This is achieved through the following ansatz:

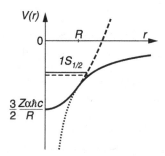

**Fig. 4.5** The effective Coulomb potential of a lead atom. At the edge of the nucleus, $r = R$, the function $r^2/R^3$, which describes the potential inside the nucleus, is matched to the hyperbola. The ground state (*solid line*) calculated from the oscillator potential lies only 1.5 MeV above the experimental result (*dashed line*)

$$V(r) = -Z\alpha\hbar c \begin{cases} \dfrac{1}{R}\left(\dfrac{3}{2} - \dfrac{1}{2}\dfrac{r^2}{R^2}\right) & r \le R \\ \dfrac{1}{r} & r \ge R. \end{cases} \qquad (4.42)$$

The $1s_{1/2}$ muon moves primarily within the nucleus and we can try to calculate the binding energy by assuming that the potential is that of a harmonic oscillator.

The Hamiltonian of the harmonic oscillator is

$$H = \frac{p^2}{2m} + \frac{m\omega^2 r^2}{2}, \qquad (4.43)$$

where, if we recall the form of the potential (4.42), we have

$$\omega^2 = \frac{Z\alpha\hbar c}{mR^3}. \qquad (4.44)$$

The ground state of the three-dimensional harmonic oscillator is at $(3/2)\hbar\omega$, so in this approximation, the binding energy is

$$\begin{aligned} E_{1s} &= -\frac{3}{2}\frac{Z\alpha\hbar c}{R} + \sqrt{\frac{Z\alpha\hbar c}{mR^3}} \\ &= -\frac{3}{2}\frac{Z\alpha\hbar c}{R}\left(1 - \sqrt{\frac{Z\alpha\hbar c}{mR}}\right) = -8.36\,\text{MeV}. \end{aligned} \qquad (4.45)$$

This is not a bad estimate. The experimental value lies a little lower because the muon is not always inside the nucleus.

## Literature

R.P. Feynman, *Quantum Electrodynamics* (Benjamin, New York, 1962)

H. Haken, H.C. Wolf, *The Physics of Atoms and Quanta* (Springer, Berlin, 2000)

V.F. Weisskopf, Search for simplicity: quantum mechanics of atoms. Am. J. Phys. **53**(3), 206–207 (1985)

# Chapter 5
# Many Electron Atoms – Shell Structure

*Necessaria est methodus ad veritatem investigandam.*
René Descartes

The most important properties of an atom are its radius and its typical excitation energy. These characterise the atom both as a molecular building block and in condensed matter.

## 5.1 Binding Energies

Similarly to the case of the hydrogen atom, we will also calculate the binding energies of complex atoms in a semiclassical approximation.

### 5.1.1 The Helium Atom

Consider two electrons in the ground state circling a helium nucleus. If we neglect the mutual repulsion of the two electrons, the average potential energy is

$$\bar{V} = -\frac{Z^2 \alpha \hbar c}{\bar{r}} = -4\frac{\alpha \hbar c}{\bar{r}} \tag{5.1}$$

and the average kinetic energy is

$$\bar{K} = 2\frac{(\hbar c)^2}{2mc^2\bar{r}^2} . \tag{5.2}$$

© Springer-Verlag GmbH Germany 2017
B. Povh and M. Rosina, *Scattering and Structures*,
Graduate Texts in Physics, DOI 10.1007/978-3-662-54515-7_5

Hence, the total energy is

$$E = -4\frac{\alpha \hbar c}{\bar{r}} + 2\frac{(\hbar c)^2}{2mc^2\bar{r}^2} \, . \tag{5.3}$$

Similar to the case of the hydrogen atom, we can calculate the binding energy and radius by minimising the energy,

$$E = \frac{4^2}{2}E_1 = -8\,\text{Ry}\,; \quad \bar{r} = \frac{2}{4}a_0 \, . \tag{5.4}$$

The experimentally determined binding energy is, though, $E = -5.8\,\text{Ry}$. The difference is clearly produced by electron–electron repulsion.

This can be well estimated by assuming that the average separation of the two electrons is $\bar{r}_{\text{eff}} = \bar{r}/0.6$. This *post hoc* assumption is justified because it delivers good results, but one can also obtain it through rather more drawn-out calculations. For us, it is important, though, that we can represent the long-range correlations between the electrons in complex atoms, which are generated by repulsion, by a single parameter for the whole periodic system of elements! The repulsive potential between the electrons is then

$$\frac{\alpha \hbar c}{\bar{r}_{\text{eff}}} = +0.6\frac{\alpha \hbar c}{\bar{r}} \, , \tag{5.5}$$

and the complete expression for the total energy is

$$E = (-4 + 0.6)\frac{\alpha \hbar c}{\bar{r}} + 2\frac{(\hbar c)^2}{2mc^2\bar{r}^2} \, . \tag{5.6}$$

The minimum energy.

$$E = \frac{(3.4)^2}{2}E_1 = -5.8\,\text{Ry} \tag{5.7}$$

agrees with measurements. The most probable radius $\bar{r}$ is $0.6\,a_0$.


## 5.1.2  Correlations

The most probable electron–electron separation in the helium atom is $r_{\text{eff}} = \bar{r}/0.6$. Does this number signify a strong or weak correlation between the two electrons? If one works out the expectation value $\langle 1/r \rangle$ with noncorrelated helium wave functions, one obtains an effective electron–electron separation $r_{\text{eff}} = \bar{r}/0.625$ (this is easy to check!). This means that the repulsion of the two electrons hardly changes their motion – a weak correlation.

### 5.1.3   The Negative H⁻ Ion

The negative H⁻ ion differs from the helium atom only through its single rather than
double nuclear charge. It is therefore more weakly bound. The binding energy of the
second electron is harder to calculate than in helium because it is very weakly bound
and one needs to be subtle to see that it is bound at all. Precise calculations yield
$E = -1.055\,\mathrm{Ry}$; because the energy of the neutral hydrogen atom is $-1\,\mathrm{Ry}$, there is
only a binding energy of $-0.055\,\mathrm{Ry} = -0.75\,\mathrm{eV}$ for the second electron. This has
been experimentally verified.

Let us try an analogous ansatz to that used for helium (5.6),

$$E = (-2 + 0.6)\frac{\alpha \hbar c}{\bar{r}} + 2\frac{(\hbar c)^2}{2mc^2 \bar{r}^2}. \tag{5.8}$$

Minimising yields a much larger radius, $\bar{r} = (1/0.7)\,a_0 = 1.43\,a_0$, than for helium
$(0.6\,a_0)$ and the energy,

$$E = -2\,(0.7)^2\,\mathrm{Ry} = -0.98\,\mathrm{Ry} > -1\,\mathrm{Ry}, \tag{5.9}$$

which is not enough for binding. One needs a tiny improvement – additional correlations between the electrons and an admixture of a configuration where the second electron is far away and polarises the remaining hydrogen atom (configuration
mixing).

The exact result can be obtained if one assumes ad hoc that the average separation
of the two electrons is $\bar{r}_{\mathrm{eff}} = \bar{r}/0.547$ instead of $\bar{r}/0.6$.

### 5.1.4   The 2s, 2p Shells

To estimate the binding energies and radii of atoms with $2 < Z \leq 10$, we will
only consider the outermost electron shell. The nucleus and the inner shells can
be described via an effective charge, $Z_{\mathrm{eff}}$. The number of electrons in the outermost
shell is correspondingly $Z_{\mathrm{eff}}$. The principal quantum number of these electrons is
$n = 2$. The potential energy of the $Z_{\mathrm{eff}}$ electrons in the Coulomb field of the $Z_{\mathrm{eff}}$
charge is

$$V = -Z_{\mathrm{eff}}^2 \frac{\alpha \hbar c}{\bar{r}}. \tag{5.10}$$

To calculate the repulsion between the electrons, we need the number of electron
pairs and their repulsive energies. The average separation between the electrons will
again be taken to be $\bar{r}_{\mathrm{eff}} = \bar{r}/0.6$. The number of pairs is

$$\frac{Z_{\mathrm{eff}}(Z_{\mathrm{eff}} - 1)}{2}, \tag{5.11}$$

and the potential energy in the shell is

$$V = \left[ -Z_{\text{eff}}^2 + 0.6 \frac{Z_{\text{eff}}(Z_{\text{eff}} - 1)}{2} \right] \frac{\alpha \hbar c}{\bar{r}} . \tag{5.12}$$

To calculate the kinetic energy, we must take into account that, for $n > 1$, quantisation of angular momentum (semiclassical orbits), $\bar{r}\bar{p} = n\hbar$, implies

$$E_{\text{kin}} = Z_{\text{eff}} n^2 \frac{(\hbar c)^2}{2mc^2 \bar{r}^2} . \tag{5.13}$$

As with the hydrogen atom, one looks for the minimum total energy. The binding energy of a closed shell with $Z_{\text{eff}}$ electrons and its radius are given by the following expressions:

$$E = -\frac{Z_{\text{eff}}\,[Z_{\text{eff}} - 0.3(Z_{\text{eff}} - 1)]^2}{n^2} \text{Ry} ,$$
$$\bar{r} = \frac{n^2}{Z_{\text{eff}} - 0.3(Z_{\text{eff}} - 1)} a_0 . \tag{5.14}$$

Using these formulae, one finds pretty good estimates for the energies and average radii – as can be seen from Table 5.1.

**Table 5.1** The most probable radius, $\bar{r}$, and the binding energy of the electrons in the outermost shells

| Element | $Z$ | $Z_{\text{eff}}$ | $n$ | $\bar{r}[a_0]$ calc. | $-E$[Ry] calc. | $\bar{r}[a_0]$ exp. | $-E$[Ry] exp. |
|---------|-----|------------------|-----|----------------------|----------------|---------------------|----------------|
| H  | 1  | 1 | 1 | 1.0 | 1.0  | 1.0 | 1.0  |
| He | 2  | 2 | 1 | 0.6 | 5.8  | 0.6 | 5.8  |
| Li | 3  | 1 | 2 | 4.0 | 0.25 | 2.8 | 0.4  |
| Be | 4  | 2 | 2 | 2.4 | 1.4  | 2.2 | 2.0  |
| B  | 5  | 3 | 2 | 1.7 | 4.3  | 1.6 | 5.2  |
| C  | 6  | 4 | 2 | 1.3 | 9.6  | 1.2 | 10.9 |
| N  | 7  | 5 | 2 | 1.1 | 18.0 | 1.0 | 19.3 |
| O  | 8  | 6 | 2 | 0.9 | 30.5 | 0.8 | 31.8 |
| F  | 9  | 7 | 2 | 0.8 | 42.0 | 0.7 | 48.5 |
| Ne | 10 | 8 | 2 | 0.7 | 69.0 | 0.6 | 70.0 |

## 5.2  Atomic Radii

The most probable radii are not easily related to measured quantities. The physically most sensible radius definition is given by $\sqrt{\langle r^2 \rangle}$. To be able to quote this, we have to know the electron density distribution.

### 5.2.1  Hydrogen and Helium

The radial wave function of the electron in the hydrogen atom's ground state is, as may be checked in any textbook,

$$R(r) = \frac{2}{\sqrt{a_0^3}}\, e^{-r/a_0} . \tag{5.15}$$

The Bohr radius, $a_0$, gives the most probable distance of the electron from the nucleus, as is easily checked if one calculates the maximum of the electron density,

$$r^2 R^2(r) = r^2 \frac{4}{a_0^3} e^{-2r/a_0}. \tag{5.16}$$

Furthermore, from (5.15), we can calculate the expectation value, $\langle 1/r \rangle$. Indeed, we find $\bar{r} = \langle 1/r \rangle^{-1} = a_0$, which explains why our estimates with $\bar{r}$ worked so well.

In scattering experiments with X-rays, the charge distribution of the atom may be measured and, from this, one can calculate the expectation value, $\langle r_H^2 \rangle$, which, for hydrogen is

$$\langle r_H^2 \rangle = \frac{4}{a_0^3} \int e^{-2r/a_0} r^4 dr = 3a_0^2 . \tag{5.17}$$

The so-defined radius of the hydrogen atom is $\sqrt{\langle r_H^2 \rangle} \approx 0.1\,\text{nm}$, and it is a better measure of the atom's size than the Bohr radius. Because the wave function of the helium atom is similar to that of hydrogen, we can estimate the size of the helium atom, $\sqrt{\langle r_{He}^2 \rangle} \approx 0.06\,\text{nm}$. Thus, the helium atom has a smaller atomic radius than hydrogen and indeed the smallest radius of all atoms.

It is not possible to calculate the radii of all noble gases from (5.14) without knowing the physical charge distributions. The radii increase slightly with charge number and are between 0.12 and 0.16 nm. We will now demonstrate, using the Thomas–Fermi model, the fact that atomic radii depend very weakly on the electron number.

## 5.2.2  Thomas–Fermi Model

A heavy atom may, *cum grano salis*, be viewed as a degenerate fermionic system. The electrons move inside a potential, $U(r)$, which is produced by the nucleus and the electrons. If the potential only slowly changes, this means that the de Broglie wavelength of the electrons only depends weakly on the radius; so one can define, for every $r$, a region $\Delta r \geq \lambda$ in which one can treat the electrons in a Fermi gas model (Fig. 5.1). The number of electrons that fit into an interval $\Delta r$ is twice the number of available phase-space cells,

$$n = \frac{2}{(2\pi\hbar)^3} \int_0^{p_F} 4\pi p^2 \mathrm{d}p \cdot 4\pi r^2 \Delta r \,. \tag{5.18}$$

From (5.18), we can easily work out the local electron density,

$$\rho(r) = \frac{n}{4\pi r^2 \Delta r} = \frac{(p_F)^3}{3\pi^2 \hbar^3} \,. \tag{5.19}$$

In this model, one assumes that the Fermi momentum, $p_F$, corresponds to the largest possible momentum of a bound electron. This is true if the electron has zero total energy, i.e., the kinetic energy is equal to the potential,

$$\frac{p_F^2}{2m_e} = eU(r) \,. \tag{5.20}$$

We now must demand that our ansatz is self-consistent: the potential $U(r)$ is determined from the charge density $-e\rho(r)$ with the help of the Poisson equation,

$$\nabla^2 U(r) = -\frac{-e\rho(r)}{\varepsilon_0} \,, \tag{5.21}$$

and the electron density is

$$\rho(r) = \frac{[2m_e eU(r)]^{3/2}}{3\pi^2 \hbar^3} \,. \tag{5.22}$$

**Fig. 5.1** The electrons in each shell of width $\Delta r$ are treated as an independent degenerate Fermi gas

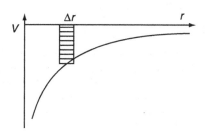

To demonstrate the scaling properties, it is helpful to rewrite (5.22) with dimensionless variables. The potential at the point $r$ is determined from the effective charge $Z_{\text{eff}}(r)$,

$$U(r) = \frac{Z_{\text{eff}}(r)e}{4\pi\varepsilon_0 r} \, . \tag{5.23}$$

Let us now introduce the variables

$$\Phi(r) = \frac{Z_{\text{eff}}(r)}{Z} \quad \text{and} \quad x = \frac{1}{(9\pi^2/2Z)^{1/3}} \frac{4r}{a_0} \approx \frac{Z^{1/3}r}{0.8853a_0} \tag{5.24}$$

so that one can write (5.21) and (5.22) in the form

$$\frac{d^2\Phi}{dx^2} = \Phi^{3/2}x^{-1/2} \tag{5.25}$$

with the boundary condition $\Phi(x \to \infty) \to 0$. This is the standard form of the Thomas–Fermi equation. It cannot be analytically solved, but it is numerically solved in Slater's book, and it is graphically represented in Fig. 5.2. This simple function reproduces a very good approximation the atomic densities, which are found from the self-consistency method. It is important that $\Phi(x)$ is a universal function that holds for all atoms ($Z \geq 10$) when one plots it as a function of $x$ in units of $0.8853a_0/Z^{1/3}$. Just as with $\Phi$, the electron densities and $\sqrt{\langle r^2 \rangle}$ in units of $0.8853a_0/Z^{1/3}$ are the same for all heavy atoms. From this, a simple scaling of the expectation values follows. For the radius, it yields $\sqrt{\langle r^2 \rangle} \propto Z^{-1/3}$. This viewpoint only holds, of course, if we consider a radius averaged over a shell. The difference between the radius of a noble gas atom and that of the succeeding alkali atom is namely larger than, e.g., the difference between the neon and xenon atoms; see Table 5.1. It may seem surprising that $\sqrt{\langle r^2 \rangle}$ decreases with $Z$, but one can easily understand it upon realising that the interior

**Fig. 5.2** Graphically represented dependence of the solution $\Phi(x)$ of the Thomas–Fermi equation on the parameter $x = (Z^{1/3}r)/(0.8853 \cdot a_0)$. *Upper right* the resulting electron density as a function of $x$

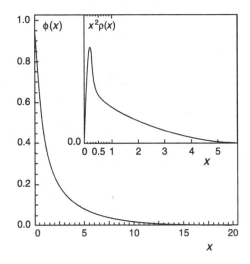

**Fig. 5.3** Comparison of the Thomas–Fermi model with the Hartree calculation for $Z = 80$

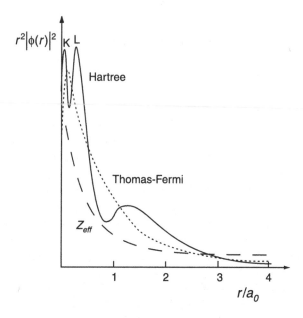

electrons as $Z$ increases lie nearer to the nucleus while the distribution of the outer electrons only slightly increases. It is therefore unsurprising that chemists, who are primarily interested in the outer electrons, use their own definitions of atomic sizes. In Fig. 5.3, a comparison of the Thomas–Fermi model with the Hartree calculation is shown. The Thomas–Fermi distribution for $x \geq 0.5$ or $r \geq 0.4 \cdot a_0/Z^{1/3}$ can be approximated by an exponential function.

### 5.2.3   Alternative Definitions

In chemistry, one uses other definitions of atomic sizes that are relevant for chemical binding.

1. The radius is defined so that the probability of finding the electron outside this radius is 50%. This definition reproduces well the separations of the atoms in covalent bonds.
2. The radius is so chosen that, at this separation from the nucleus, Pauli repulsion does not let another atom approach more closely. This definition is used for the separations of atoms in ionic bonds.

The different definitions lead to systematically varying values but reproduce well the general dependence of atomic sizes on $Z$.

## 5.3 Atoms with Magnetic Moment

In the hydrogen atom, states with the same principal quantum number, $n$, but different angular momenta, $\ell$, are degenerate. They all have the same kinetic and potential energy. This means that they all have the same $\langle 1/r \rangle$ and thus the $\langle r \rangle$ values are very similar. This does not mean that they all extend equally. It should not be forgotten that, for example, the 3s state has two, the 3p state has one and the 3d state no radial nodes. Thus, the only 3d maximum is hidden in the external maxima of the 3s and 3p states (Fig. 5.4). This effect is still stronger for heavier atoms. As a first example, consider atoms with incomplete d subshells; typical examples are iron, cobalt and nickel. In these elements, the electrons in the inner maxima of the s and p states feel the almost completely unscreened Coulomb potential of the nucleus, while the only maximum of the 3d state lies in the middle of the electron shells and so experiences a strongly screened potential. Thus, the 3d state has a higher energy than the 3s and 3p states and may be better compared with the 4s and 4p states. The periodic table of the elements makes this clear, as the electrons occupy the levels in the following order: 1s; 2s, 2p; 3s, 3p; 4s, (3d, 4p); .... .

The s and p electrons in an atom with an incompletely occupied outermost shell are "chemically very active". Thus, the states of neighbouring atoms combine in covalent or ionic bonds in such a way that the external shells are effectively filled up. It follows from this that the s and p electrons in most stable molecules are pairwise coupled with opposite angular momenta. Thus, the sum of their magnetic momenta is equal to zero and the substance is diamagnetic and only has an induced magnetisation.

It is very different for an incomplete d subshell. Spins and orbital moments of the electrons can orient themselves in a parallel fashion, and their magnetic moments add up. Therefore substances of such atoms are paramagnetic. Other systems of such atoms (crystals) can even be ferromagnetic.

For paramagnetism, it is crucial that the d states, which have a higher energy than the corresponding s and p states and are later occupied, lie, geometrically speaking, deeper inside the atom than the s and p states and are thus protected from chemical influences. Thus, they can afford to remain unpaired. Even when several electrons are in the d subshell, it is energetically preferred for the electrons to line up in parallel.

**Fig. 5.4** Electron densities in the $n = 3$ shell with an incomplete 3d subshell

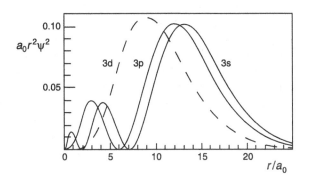

For a symmetric spin function, the spatial wave function is antisymmetric and thus the Coulomb repulsion is minimised. This is not a magnetic but rather an electrostatic effect.

Similar consideration also holds for those rare earth atoms with unpaired electrons in the f subshell. These lie above the energies of the corresponding electrons in the d subshell and are geometrically still more screened by the outermost s, p and d electrons. Therefore, the rare earths (e.g., samarium, europium) with partially occupied f subshells are even better (if more expensive) ferromagnets than iron.

## 5.4  Ferromagnetism and Antiferromagnetism

Ferromagnetism is the example *par excellence* of a phase transition and is often used as a model in other areas of physics. Therefore, we will briefly sketch this transition. The phenomenon of ferromagnetism is a consequence of the lattice structure. The s and p electrons, in the case of iron, and the s, p and d electrons, for the rare earths, are involved in the binding of the crystal lattice, the d or f electrons are screened and barely overlap with neighbouring atoms; however, this is enough for the total wave function of the d or f electrons to have to be antisymmetric. For ferromagnetism, it is energetically preferred that the angular momenta of the d electrons are parallel to each other and thus the spatial wave function is antisymmetric. This lessens the Coulombic repulsion energy: an antisymmetric wave function has a node where two electrons overlap and so the Coulomb energy is minimised.

For antiferromagnetic substances, the situation is reversed. An antisymmetric spin wave function and a symmetric spatial wave function increase the Coulomb attraction between neighbouring ions, and it is larger than the repulsion of the electrons.

Let us estimate the binding energy of the d electrons, which is responsible for the phase transition between paramagnetic and ferromagnetic states. The Curie point of iron lies at $T_C \approx 1000\,\mathrm{K}$, which corresponds to a binding energy of around $0.1\,\mathrm{eV}$. Magnetisation is the best indicator of the orientation of the magnetic moments of the electrons, and in the paramagnetic state, it is well described by the Curie law,

$$\chi_P = \frac{C}{T}. \tag{5.26}$$

Here, $\chi_P$ is the paramagnetic susceptibility and $C$ a material-dependent constant. The magnetisation, $\mu_0 M$, is the consequence of an external magnetic field, $B_a$, and of the electrostatic interaction of the electrons in the lattice, which may be formally parameterised as an effective field, $B_e = \lambda M$,

$$\mu_0 M = \chi_P (B_a + B_e) = \chi_P (B_a + \lambda M). \tag{5.27}$$

Here, $\lambda$ is a phenomenological constant.

The ansatz (5.27) is typical for the formulation of a phase transition in which the critical temperature of the phase transition is determined by the interaction between the constituents of the system. We will use similar ansatz in other cases. The ansatz (5.27) contains a positive feedback for the magnetisation. The quantity that measures the degree of order in the phase transition, here the magnetisation, is called the order parameter.

Taking the terms involving the magnetisation to the left-hand side and using the Curie law (5.26), one finds

$$\mu_0 M = \frac{C}{T - T_C} B_a, \qquad T > T_C. \qquad (5.28)$$

The pole at the temperature $T_C = C\lambda/\mu_0$ signals the phase transition. Naturally, the magnetisation cannot increase beyond the saturation value. Near and below $T_C$, one must apply the improved Curie law, which takes saturation nontrivially into account. Then (5.27) is no longer linear and, at $T > T_C$, is already fulfilled nontrivially for $B_a = 0$.

## Literature

N.W. Ashkroft, N.D. Mermin, *Solid State Physics* (Holt, Rinehart and Winston, New York, 1976)

H. Haken, H.C. Wolf, *The Physics of Atoms and Quanta* (Springer, Berlin, 2000)

C. Kittel, *Introduction to Solid State Physics* (Wiley, New York)

J.C. Slater, *Quantum Theorz of Atomic Structure* (McGraw-Hill, New York, 1960)

V.F. Weisskopf, Search for simplicity: atoms with several electrons. Am. J. Phys. **53**(4), 304–305 (1985)

# Chapter 6
# Covalent and Ionic Binding – Electron Sharing

*Durch das Einfache geht der Eingang zur Wahrheit.*

Lichtenberg

The binding energies in atoms were determined through the condition that the total energy of an isolated atom, the sum of the potential and kinetic energies of the electrons, has a minimum. In interactions with other atoms, they form complex structures such as molecules, glasses or crystals. Charge polarisation of the external electrons causes the total energy of the molecules to be lower than the sum of the energies of the isolated atoms. In this chapter, we only consider the chemical binding that leads to a compact molecular or crystalline structure. This can be approximately formed from two simple idealisations of bonds: covalent and ionic bonds. The metallic bond is a *delocalised* covalent bond. We treat this in Chap. 11 as an example of a degenerate fermionic system.

## 6.1 The Covalent Bond

The ideal example of a purely covalent bond is the hydrogen molecule. This example is very attractive because one can almost sketch it *on the back of an envelope*. In our qualitative considerations, we use molecular orbitals to show that the covalent bond is a purely electrostatic affair and not an exchange phenomenon, as the alternative viewpoint with atomic orbitals suggests. The term *orbital* is often used in atomic physics and chemistry to mean a single-particle wave function

In textbooks, atomic orbitals are usually used as a basis. The two separate hydrogen atoms are brought toward each other and, as the electrons of the two atoms begin to overlap, the composite wave function is formed. To describe the symmetric spatial wave function, one uses exchange coordinates, which have no physical significance.

© Springer-Verlag GmbH Germany 2017
B. Povh and M. Rosina, *Scattering and Structures*,
Graduate Texts in Physics, DOI 10.1007/978-3-662-54515-7_6

The electrons are only exchanged in the sense that, in the molecular quantum state, it is no longer possible to assign the electrons to individual protons.

In our derivation, we begin with a helium atom, the nucleus of which splits into two deuterons, and imagine what would happen to the electron cloud. We, though, do not use deuteronic but proton masses because the covalent bonds in both $^1H_2$ and $^2H_2$ are very similar.

### 6.1.1  The Hydrogen Molecule – A Case of Broken Symmetry

Both electrons in the ground state of the molecule are coupled with total spin $S = 0$, their spatial wave function is symmetric and predominantly corresponds to a molecular orbital with orbital angular momentum $L = 0$. Here, we will show that, in the case of the hydrogen molecule, the molecular orbitals quickly produce the correct result. The primary attraction is caused by a helium atom-like charge distribution of the electrons around the two hydrogen nuclei.

Let the separation between the two protons be $d$ and the separation of the electron from the molecular centre of mass be $r$ (Fig. 6.1). For $d/2 \ll r$, the total energy of the electrons is equal to that in helium. For $d/2 \gg r$, one has to take into account that the electrons mostly each only see one proton and the total energy is that of two separated hydrogen atoms. We will simulate the connection between the two regions via the following ansatz for the total energy of the hydrogen molecule:

$$E = 2\frac{\bar{p}^2}{2m} - 2\frac{\alpha\hbar c}{\bar{r}}\left[1 + \left(1 - e^{-2\bar{r}^2/d^2}\right)\right]$$
$$+ 0.6\frac{\alpha\hbar c}{\bar{r}}\left(1 - e^{-2\bar{r}^2/d^2}\right) + \frac{\alpha\hbar c}{d}. \tag{6.1}$$

The last term accounts for the contribution to the total energy from the mutual proton repulsion, the first three terms correspond to the electron contributions. For $\bar{r} \gg d/2$, this contribution to (6.1) is equal to that in helium (see (5.6)), while for $\bar{r} \ll d/2$, it is equal to that in two separated hydrogen atoms.

**Fig. 6.1** The separation between the two protons is $d$, the separation of the electron from the molecule's centre of mass is $r$; the contour of the hydrogen molecule at a radius $r \approx 2a_0$ has the value $0.001$ electrons/$a_0^3$

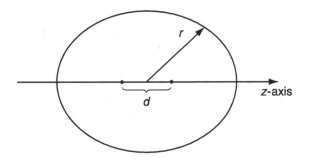

We will see that the electrons are mostly distributed around $\bar{r} > d/2$ and we may – as in the case of the helium atom – take $\bar{r}\bar{p} = \hbar$. If we re-express $\bar{r}$ and $d$ in terms of the Bohr radius and let $\xi = \bar{r}/a_0$ and $\eta = d/2a_0$, the equation becomes

$$E = \left\{ \frac{2}{\xi^2} - \frac{4}{\xi}\left[1 + 0.7\left(1 - e^{-\xi^2/2\eta^2}\right)\right] + \frac{1}{\eta} \right\} Ry, \qquad (6.2)$$

where Ry is the Rydberg constant (4.4).

Figure 6.2 displays the total energy, i.e., the sum of the electron attraction, $E'$, and mutual proton repulsion. The minimum lies at $d \approx a_0$, and the most probable electron radius in the molecule is $\bar{r} = 0.9a_0$. The resulting binding energy is $E_{bind} = E + 2\,Ry = -0.47\,Ry$. These values should be compared with the experimental values $d = 1.43a_0$ and $E_{bind} = -0.34$ Ry.

This result shows that the assumption $\bar{r} > d/2$ is justified and that the electron distribution is similar to the helium atom. It should be stressed again that, for the distribution of the electrons, the effective radius, $\sqrt{\langle r^2 \rangle}$, is relevant and is around 1.7 times larger than the most probable radius.

The question remains, though, how good the assumption of a spherically symmetric charge distribution is. The two charge centres destroy spherical symmetry. This can be best tested by considering the rotational states of the molecules. We should note that the two proton spins can be parallel (*orthohydrogen*) or antiparallel (parahydrogen). Since the two-proton wavefunction must be antisymmetric, the orthohydrogen can have only odd two-proton orbital angular momenta and parahydrogen can have only even ones.

The magnetic moments of the rotational states are formed from the magnetic moments of the rotating protons and – with opposite signs – of the electrons. In the first excited state ($J = 2$) of the hydrogen molecule the electron spins as well as the

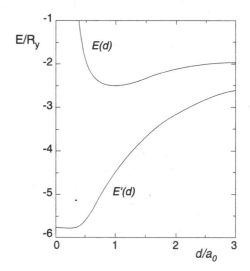

**Fig. 6.2** The energy of the two hydrogen atoms as a function of the separation $d$ between the nuclei. $E'$ signifies the binding energy of the electrons to the two hydrogen atoms as a function of the separation $d$. For $d = 0$, we obtain the binding energy of the helium atom, $-5.8$ Ry, while as $d \to \infty$, we find the binding energy of the two hydrogen atoms, $-2$ Ry. The total energy $E$ is found by adding the mutual repulsion of the two nuclei to $E$

**Fig. 6.3** Results of an exact calculation of the electron distribution in the hydrogen molecule. The contours correspond to electron densities (*from outside to inside*) of 0.0010, 0.0025, 0.0050, 0.01, 0.025, 0.05, 0.10 and 0.25 electrons/$a_0^3$

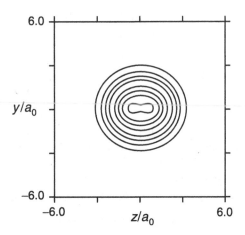

proton spins are antiparallel and their spin magnetic moments do not contribute; only orbital currents contribute to the magnetic moment. The measured magnetic moment in this state is $\mu_{H_2} = (0.88291 \pm 0.0007)\mu_N$. Here, $\mu_N$ is the nuclear magneton. Two protons that rotate around their centre of mass with angular momentum, $\hbar$, generate a magnetic moment of the size of the nuclear magneton. The 12% smaller result indicates a contribution, if a small one, from the electrons to the magnetic moment. Electrons with $S = 0$ and $L = 0$ do not contribute to the rotation. The contribution comes from electrons with $L = 2$. This means that, as well as the spherically symmetric electron distribution, there is also a quadrupole one. From the experimental values of the magnetic moment, the quadrupole moment ($Q = \langle 3z^2 - r^2 \rangle = 0.59a_0^2$) and the mean square radius of the electrons in the hydrogen molecule ($\langle r^2 \rangle = 2.59a_0^2$), it follows that the probability of finding the electron in the $L = 2$ state is roughly 20%.

In Fig. 6.3, the electron densities in the hydrogen molecule are sketched. The chemical bond is an electrostatic effect: the electrons taking part in the bond feel twice the charge of an individual atom. This attraction is greater than the repulsion of the two protons.

### 6.1.2  An Analogy

Let us draw the potential in the hydrogen molecule in space (see Fig. 6.4). Because of the fact that spherical symmetry is broken, there are two new excitation modes: rotation around symmetry axis, through an angle $\phi$, and radial oscillations orthogonal to the eaves of the potential. The potentials in the analogous cases of chiral symmetry (Fig. 12.9) and the Higgs field (Fig. 16.10) indeed show a similarity to the hydrogen molecule potential. One would not, though, tend to call this a Mexican hat potential, but rather as a witch's hat potential. The rotation and vibration in the cases of chiral

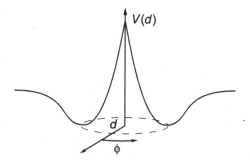

**Fig. 6.4** The dependence of the hydrogen molecule potential on the interatomic separation $d$. The range of the nuclear force is greatly exaggerated. The separation $d$ at which the Coulombic repulsion is compensated by the nuclear force is exaggerated. The angle $\phi$ signifies the rotation angle. Vibrational states correspond to movement of the atoms in the radial direction

symmetry and the Higgs field correspond to quantised waves – mesons and Higgs bosons, respectively – because those cases involve infinitely many coupled degrees of freedom.

### 6.1.3 Covalent Bond in the (2s, 2p) Shells

Our treatment of $H_2$ can be extended to other symmetric diatomic molecules, $Li_2$, $N_2$, $O_2$, .... The quadrupole part of the electron cloud is greater in heavy molecules than in light ones because the separations between the atoms are greater due to repulsion effects.

The covalent bond in carbon, which generates the immense variety of organic molecules, is especially interesting. It is impressive that, in the cases of C–H, C–C, C=O, the bond energies of each of these covalent bonds only vary by maximally 10%; they are around 4.5 eV. The same value, up to 10%, holds for the H–H bonds in the hydrogen molecule and O–H in water molecules. Clearly, the molecular orbitals in all these cases can be well described in terms of atomic wave functions. This is done by using a superposition of 2s and 2p states (so-called hybridisation). Hybridisation yields the characteristic angles between the bonds.

### 6.1.4 Carbon - The Magic Atom

Carbon plays a fundamental role for the life on the Earth. With its valence bond it can form a tremendous variety of organic compounds and different crystal structures.

Carbon has 2 electrons in the 1s shell and 4 electrons in the 2s2p shell. Since the 2s and 2p subshells are almost degenerate the so called *hybride orbitals* (superpositions of 2s and 2p orbitals) are easily formed and they offer four strong covalent bonds. Since one 2s and three 2p orbitals participate the hybrid configuration is denoted as $sp^3$.

In **organic compounds** a chain or ring of carbon atoms is formed, with extra bonds connected to different atoms (H, Cl, N....).

In **diamond**, a cubic lattice is formed with the four bonds connected to four neighbouring carbon atoms, at an angle of 109.5° (Fig. 12.13). Because of the strong covalent bonds diamond is known to be the hardest crystal. The excitation energy of electrons in next shells is about 6 eV, therefore diamond is transparent for optical frequencies. Also, there are practically no electrons in the higher shells, therefore diamond is an electric insulator and poor thermal conductor. It should be noted that silicon and germanium have also four valence bonds, they have a similar crystal structure as diamond, but are good semiconductors (the gap between the valence and higher shell is only 1.1 eV for Si and 0.67 for Ge).

In a way, it is surprising that carbon can form also a two-dimensional lattice called **graphene** (Fig. 6.5a). Only three electrons participate in the covalent bond

**Fig. 6.5**  Graphene, Graphite, Fullerene

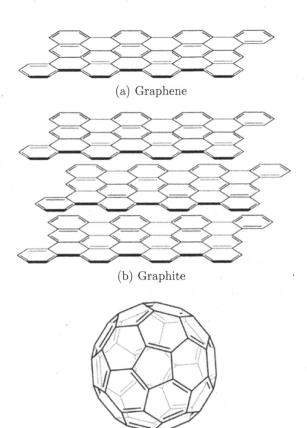

(a) Graphene

(b) Graphite

(c) Fullerene

(the $sp^2$ hybrid configuration with bonds at $120°$). However, the fourth electron is not localized and forms a bond similar to metal bond with neighbouring atoms. Therefore graphene is a good electric conductor. Due to strong covalent bond it is the strongest material we know, 300 times stronger than steel. As a monoatomic sheet, it has promising applications as support in microscopy, lubricant, protective layer and many more. Its structure is sometimes compared to aromatic organic compounds with the fourth bond acting as a delocalized double bond – an infinite extension of benzene, naphthalene, anthracene...

The most stable form of carbon, the **graphite** can be visualized as a stack of graphene layers loosely bound by the Van der Waals force (Fig. 6.5b). The sum of the three covalent bonds, the "metalic" bond and the Van der Waals binding energy suffices to make the enthalpy of graphite at room temperature and pressure 0.031 eV below diamond. However, at higher pressures (densities) diamond is more economical than graphite. Diamond naturally forms at pressures 4.5–6 GPa and temperatures 900–1300 °C. Due to free electrons, graphite is a good electric conductor. Due to the weak Van der Waals force, the layers are far apart and easy to remove; graphite is brittle, a good lubricant, it leaves black marks - it name derives from the Greek word *graphein* - to write.

Interesting two-dimensional structures are also **nanotubes** - piece of graphene wrapped into a cylinder - and **fullerenes** - pieces of graphene wrapped into a sphere (Fig. 6.5c). They both find applications in electronics, pharmacy, lubricants and protectives; new proposals are rapidly developing. While graphene and nanotubes consist of hexagons, the fullerene must have 12 pentagons (in addition to hexagones) due to the Euler's theorem n(vertices)+n(faces)−n(edges)=2, for simple topology. In fact, for *buckminster fullerene* $C_{60}$ (named after the dome of the architect Buckinster Fuller) 60+(12+20)−90=2. There are also larger fullerene structures. Nanotubes and fullerenes appear spontaneously in soot in small quantities and were only discovered in last decades. Nowadays they can be produced copiously in laboratory with appropriate treatment of "soot".

### 6.1.5 Energy Source Oxygen

The oxygen double bond (O=O) is a covalent bond. The bond energy is, though, only as large as a single bond energy for light atoms in the (2s, 2p) shells and in the hydrogen molecule.

The oxygen atom has an almost full shell, but only two electrons that participate in bonds. The rest repel atoms due to the Pauli principle. This leads to a greater interatomic separation, so that the common orbitals have a smaller overlap. The bonding energy in the oxygen molecule is reduced by a factor of two compared with the above-mentioned compounds.

The $O_2$ molecule, with its double bond, thus has only around 2.3 eV per bond. This is why, when one burns carbon, hydrogen, or other compounds together with oxygen, one gains around 2.2 eV per bond. This is why oxygen is so chemically active and mostly found in chemical compounds. Atmospheric oxygen is continually resupplied as a by-product of photosynthesis.

One speaks of fossil fuel energy sources and means coal, gas and oil. The energy is, though, stored in atmospheric oxygen! Let us consider the burning of methane with oxygen:

$$CH_4 + 2O_2 \rightarrow CO_2 + 2H_2O. \tag{6.3}$$

The number of covalent bonds remains constant in the reaction, as the four weak oxygen bonds are replaced by four stronger ones. Photosynthesis, which separates oxygen from carbon, has stored the energy in the weak bond in oxygen.

## 6.2  Ionic Bonds

Typical examples of such bonds are LiF, NaCl, CsI, …. Comparison with experiments (electric dipole moments of molecules) shows that the electrons of the alkali atoms are up to 90% transferred to the halogen atom. Both ions thus have a noble gas-like closed shell. We will now assume that the electron is completely transferred. The two ions attract each other until a further overlap of the electron clouds is stopped by the Pauli principle. For a NaCl molecule, this takes place at a separation around $d = 0.24$ nm. (In a crystal (see Fig. 1.6), the separation is a little larger, $d = 0.28$ nm.) The binding energy of the molecule relative to free ions is then

$$E - E_{ions} = -\frac{\alpha \hbar c}{d} = -\frac{2Ry}{d/a_0} = -5.6\,eV. \tag{6.4}$$

More important than this number is the bond energy compared with neutral atoms. To take an electron from an alkali atom and give it to a halogen atom requires $1.5\,eV$. This implies a bond energy in NaCl of

$$E - E_{atoms} = -4.1\,eV. \tag{6.5}$$

Ionic molecules are mostly found built into crystals. The ionic charge is not screened and the long-range Coulomb force must be taken into account. In a crystal, the binding energy per atom is reduced to around 78% of its former value, i.e., not only the attraction by immediate neighbours, which have the opposing charge, but also interactions with more distant atoms, both of the same and the opposite charge, are significant.

# Literature

N.F. Ramsey, *Molecular Beams* (Oxford University Press, New York, 1958)

V.F. Weisskopf, Search for simplicity: chemical energy. Am. J. Phys. **53**(6), 522–523 (1985)

V.F. Weisskopf, Search for simplicity: the molecular bond. Am. J. Phys. **53**(5), 399–400 (1985)

# Chapter 7
# Intermolecular Forces – Building Complex Structures

*Pluritas non est ponenda sine necessitate.*

William of Occam (Ockham)

## 7.1 Van der Waals Interaction

Neutral atoms and molecules may be pictured as rapidly oscillating dipoles with frequencies of the order of $\hbar\omega_0 \approx \alpha\hbar c/2a_0$ and dipole moment sizes $\mu_{\text{el}} \approx ea_0$. A classical spherically symmetric charge distribution does not produce a dipole moment, but in quantum mechanics, the uncertainty in the electrons' coordinates (see (4.2)) produces one.

Temporal correlations in the dipole moments generate van der Waals forces between atoms and molecules. These forces only play a dominant role when other varieties of chemical bonds are not present. This is true between noble gas atoms, between molecules in organic compounds and, e.g., also holds for the bonding between the crystal layers in graphite, which are themselves constructed through covalent bonds. The following brief treatment of the van der Waals force uses the hydrogen atom to compare its scale to known atomic constants. The covalent bond in hydrogen is certainly stronger than the contribution of the van der Waals interaction. The order of magnitude that we obtain from this hypothetical case is a very good estimate of the van der Waals force between hydrogen molecules.

We treat the van der Waals force in some detail as, at present, experiments to do with the Casimir effect are fashionable. This tests the effect of vacuum fluctuations at macroscopic scales.

© Springer-Verlag GmbH Germany 2017
B. Povh and M. Rosina, *Scattering and Structures*,
Graduate Texts in Physics, DOI 10.1007/978-3-662-54515-7_7

**Fig. 7.1** Atom in front of an
ideal conducting wall and its
virtual image

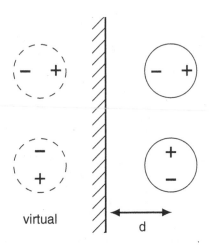

### 7.1.1  Van der Waals Interaction Between an Atom and a Conducting Wall

An atom next to an ideal conducting wall generates a mirror charge (Fig. 7.1), which oscillates exactly in time with the atom. This quasi-static approximation only holds for separations $a_0 < d < a_0/\alpha$. The upper limit is because, for large separations, the atom and its image get out of phase and one must take retardation effects into account. In the quasi-static approximation, the potential energy of the oscillating dipole $a$ in front of our ideal conducting wall is

$$V_{a,w}(R) \approx -\alpha\hbar c \frac{a_0^2}{(2d)^3} . \tag{7.1}$$

### 7.1.2  Van der Waals Interaction Between Two Atoms

The oscillating dipoles of two atoms are not correlated right from the start. The correlations are produced by communication between the atoms through so-called two-photon exchange. The van der Waals interaction may be estimated as follows: the binding energy of one dipole in the field of the other is proportional to $V_{a,w}$ (see (7.1)). The excitation energies of the atoms in the presence of a dipole moment have a typical value of $\Delta E \approx 1$ Ry. At large enough separations $R$ the second-order perturbation formula is valid and the van der Waals (atom–atom) interaction is

$$V_{a,a}(R) \approx -\left(\alpha\hbar c \frac{a_0^2}{R^3}\right)^2 \frac{1}{2\Delta E} \approx -\alpha\hbar c \frac{a_0^5}{R^6} . \tag{7.2}$$

For separations $R > a_0/\alpha$, the time $R/c$ taken by a signal between the atoms is larger than the typical oscillation time, $\hbar/\text{Ry} = a_0/(c\alpha)$. One must take retardation into account and, as one would expect, the interaction falls off faster than $1/R^6$. To estimate the van der Waals interaction for separations $R > a_0/\alpha$, a different viewpoint separation $R$ is helpful. This so-called Casimir effect has become especially interesting due to new measurements of the forces between neutral conducting surfaces, and we will also briefly consider this phenomenon.

### 7.1.3 Van der Waals Interaction and the Casimir Effect

To take retardation into account, let us consider the dipole oscillations to be a consequence of the zero-point oscillations of the electromagnetic field. The fluctuating electric fields (see Sect. 4.2.1) induce dipole moments in neutral systems, which contribute to the van der Waals interaction.

Let us consider two neutral, but polarisable, objects 1 and 2 separated by a distance $R$. The fluctuating electric field, $\mathcal{E}(\mathbf{r}, t)$, of the zero-point energy polarises both objects and gives them electric dipole moments

$$\begin{aligned} \mu_1 &= \varepsilon_0 \alpha_1 \mathcal{E}(\mathbf{r}_1, t) \\ \mu_2 &= \varepsilon_0 \alpha_2 \mathcal{E}(\mathbf{r}_2, t), \end{aligned} \qquad (7.3)$$

where $\alpha_1$ and $\alpha_2$ are the polarisabilities of the two objects and $\varepsilon_0$ is the dielectric constant. In the following derivation, we will ignore the angular dependence. The energy of dipole 1 in the radiation field of dipole 2, $\mathcal{E}_2$, is

$$W = -\mu_1 \mathcal{E}_2(r_1, t). \qquad (7.4)$$

The radiation field of a Hertzian dipole of size $\mu_2$ oscillating with frequency $\omega$ is well known to be

$$\mathcal{E}_2 = \frac{1}{4\pi\varepsilon_0} \mu_2 \frac{\omega^2}{c^2 R}. \qquad (7.5)$$

The contribution to the binding energy of the zero-point oscillations with frequency $\omega$ follows from (7.4) and (7.5)

$$W = -\frac{\varepsilon_0}{4\pi} \alpha_1 \alpha_2 \mathcal{E}_\omega(\mathbf{r}_1, t) \mathcal{E}_\omega(\mathbf{r}_2, t) \frac{\omega^2}{c^2 R}. \qquad (7.6)$$

Here, the electric fields $\mathcal{E}_\omega(\mathbf{r}_{1,2}, t)$ are the Fourier components of $\mathcal{E}_{1,2}$.

The total binding energy is found by integrating (7.6) over the phase space of the oscillations of the electromagnetic field,

$$W = -\frac{\varepsilon_0}{4\pi} \int \alpha_1 \alpha_2 \mathcal{E}_\omega(\mathbf{r}_1, t) \mathcal{E}_\omega(\mathbf{r}_2, t) \frac{\omega^2}{c^2 R} \frac{L^3 4\pi \omega^2 d\omega}{(2\pi c)^3} . \tag{7.7}$$

Here $L$ is the size of the normalisation box and cancels in the final result. The upper integration limit is $\omega \approx c/R$. This is because, for frequencies $\omega \gg c/R$, the product $\mathcal{E}_\omega(\mathbf{r}_1, t)\mathcal{E}_\omega(\mathbf{r}_2, t)$, viewed as a function of $\omega$, oscillates very rapidly and does not contribute significantly to the integral. For frequencies $\omega \ll c/R$, the product may be taken to be constant, and thus, the average energy of the vacuum fluctuations is

$$L^3 \varepsilon_0 \mathcal{E}_\omega(\mathbf{r}_1, t)\mathcal{E}_\omega(\mathbf{r}_2, t) \approx L^3 \varepsilon_0 \mathcal{E}^2 \approx \hbar\omega . \tag{7.8}$$

This implies the following contribution of the zero-point fluctuations to the van der Waals interaction

$$W \approx - \int_0^{c/R} \alpha_1 \alpha_2 \frac{\hbar\omega^5}{c^5 R} d\omega . \tag{7.9}$$

Here, we have dropped the prefactors because our approximations in the integration were so rough that it would not be appropriate to give the impression that the upper bound was more than an order-of-magnitude estimate. Here, we denote the van der Waals interaction by $W$ (instead of $V$, as in (7.2)) because we are here employing a different point of view (the energy of the electromagnetic field).

The polarisability of the hydrogen atom is $\alpha_H \approx a_0^3$. Other atoms also have polarisabilities on this scale; so the contribution of zero-point fluctuations to the interaction between two atoms is

$$W_{a,a} \approx -\hbar c \frac{a_0^6}{R^7} . \tag{7.10}$$

This approximation was derived under the assumption that the atoms undergo forced vibrations in the field of the zero-point oscillation and thus only holds for $R > a_0/\alpha$. At smaller separations, the contributing frequencies $\omega \approx c/R$ are larger than typical atomic frequencies, $Ry/\hbar \approx \alpha c/a_0$, so the atoms cannot follow the forced vibration. Thus, at a separation $R \approx a_0/\alpha$, the mechanism that generates synchronous dipole oscillations and, hence, the van der Waals interaction, changes. For $R < a_0/\alpha$, the zero-point oscillations of the atoms can mutually synchronise each other, while, for $R > a_0/\alpha$, the synchronisation takes effect through the common external influence of the zero-point oscillations of the radiation field.

### 7.1.4 Wall–Wall Interaction

As we saw in Sect. 4.2 about the Lamb shift (see (4.24)) $u_L$, the energy density of the zero-point fluctuations of the electromagnetic field in a sufficiently large volume

$L^3$ may be calculated to be

$$u_L = \int\limits_0^{\omega_{max}} 2\frac{\hbar\omega}{2}\frac{4\pi\omega^2 d\omega}{(2\pi c)^3} = \frac{\hbar\omega_{max}^4}{8\pi^2 c^3}. \tag{7.11}$$

If we place a plate capacitor with perfectly conducting walls into this sufficiently large volume, the so-called Casimir force acts on the walls. Let $S$ be the size of the areas of the plate capacitor and $d$ be the separation of the plates. Only those fluctuations with nodes at the walls are possible in the capacitor. The lowest frequency of the fluctuations corresponds to the wavelength $\lambda = 2d$ which implies $\omega = \pi c/d$. The energy density $u_K$ in the capacitor is the difference between the energy density $u_L$ in the volume $L^3$ and the sum of the fluctuations that are excluded by the boundary conditions.

$$u_K \approx \int\limits_{\pi c/d}^{\omega_{max}} \hbar\omega\frac{4\pi\omega^2 d\omega}{(2\pi c)^3} = \frac{\hbar\omega_{max}^4}{8\pi^2 c^3} - \frac{\pi^2\hbar c}{8d^4}. \tag{7.12}$$

This calculation, with a sharp cut off at $\omega_{min}$, is not exact; one ought to calculate the discrete sum of the zero-point fluctuations in the surviving oscillatory modes. Because the calculation is somewhat tedious, we here state its result: the 8 in the denominator of the last term of (7.12) must be replaced by 720. The difference between the energy densities outside and inside the capacitor is thus obviously

$$\Delta u = u_K - u_L = -\frac{\pi^2\hbar c}{720d^4}. \tag{7.13}$$

From the difference of the two energy densities, we can calculate the pressure on the plates:

$$P_{Casimir} = -\frac{1}{S}\frac{d(\Delta u S d)}{dd} = -\frac{\pi^2\hbar c}{240d^4}. \tag{7.14}$$

The Casimir force has indeed been experimentally confirmed in various capacitor geometries and at separations on the scale of $\mu$m. It is thus believed that the existence of zero-point fluctuations has been demonstrated for macroscopic scales. An extrapolation of the Casimir effect to astronomical dimensions leads, though, to absurd results: to energy densities that are orders of magnitude larger than the energy densities found in modern experiments.

   Let us now show that the formula for the Casimir force (7.14) also follows from the expression (7.10) for the retarded van der Waals force. The walls are now made from dielectric atoms with polarisability $\alpha_H = a_0^3$. For this estimate, we only consider the atoms in two blocks with area $S = d^2$ and depth $d/2$ (Fig. 7.2). We neglect contributions to the force from atoms outside these regions. The number of atoms in each cube is $N = \frac{1}{2}(d/2a_0)^3$, and one so obtains, from (7.10),

**Fig. 7.2** Wall–wall interaction as the sum of the retarded atom–atom interactions

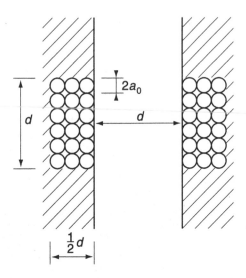

$$P_{\text{Casimir}} \approx -\frac{N^2}{S}\frac{dW_{\text{aa}}}{dR} \approx -\frac{7}{256}\frac{\hbar c}{d^4}. \tag{7.15}$$

Here, we have assumed $R \approx d$. The agreement is not so bad for our rough approximation.

## 7.2 Hydrogen Bridge Bond

A special bond between molecules is produced when two molecules share a hydrogen nucleus between them. Without its electron, a hydrogen atom is a "naked" proton, a tiny object five orders of magnitude smaller than an atom. This gives the hydrogen atom a special status in chemistry and makes possible a special sort of bond, the hydrogen bridge bond. This situation occurs when the energy of a proton between two atoms has two minima. In this case, the wave function of the proton is a superposition of the wave functions centred around each of the minima. The best known example is the bond between water molecules, which is responsible for the exotic behaviour of water. The spatial construction of biologically active molecules is also made possible by the hydrogen bridge bond.

### 7.2.1 Water

Water has three noteworthy properties that are crucial for life and for the environment. Liquid water ($<10\,°C$) is heavier than ice, has an exceptionally large specific heat and is – because of its large dipole moment – an excellent solvent.

**Fig. 7.3** The electron distribution in a water molecule. The contours correspond to relative electron densities of 0.10, 0.17 and 0.30

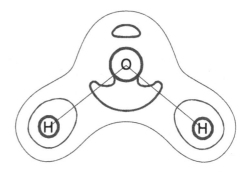

## 7.2.2 Water Molecule

All of the above quantities follow from the structure of the water molecule. The two covalent bonds, H–O–H, are at an angle of 104.5°. It is energetically favoured for the molecular orbitals to overlap strongly with the atomic orbitals of the valence electrons. Two orthogonal 2p orbitals are maximally correlated at an angle of 90°. A superposition of 2s-2p may be at any angle between 90° and 120°, and an admixture of 2 s is energetically less favoured. Hybrid orbitals at an angle of 104.5° (Fig. 7.3) optimise the Coulombic attraction of the electrons to the protons and the Coulombic repulsion between the two protons. The electron distribution has its charge centre nearer to the oxygen than to the two protons (Fig. 7.3). The consequence is a sizeable electric dipole moment ($\mu_e = 0.068\, e\, a_0$).

## 7.2.3 Model of the Hydrogen Bridge Bond

Let us consider a proton in a covalent bond with oxygen. When a second oxygen atom approaches the proton, the proton sees a potential with two minima (Fig. 7.4) and tunnels through the potential barrier from one minimum to the other. This produces an energy shift, which we will now roughly estimate.

The proton is bound to the oxygen atom by a harmonic oscillator potential (Fig. 7.4 left). A typical vibrational energy of the proton in an isolated water molecule is $\Delta E_{vib} \approx 0.3\,\text{eV}$. The vibrational ground state has then an energy of $\approx 0.15\,\text{eV}$. When, though, the proton feels the attraction of two oxygen atoms, then the potential that the proton moves in is broader than the individual potentials (Fig. 7.4 right). The vibrational energy of the proton in the new ground state is smaller by roughly a factor of two. This implies the correct order of magnitude of the hydrogen bridge bond which corresponds to the difference in the energies of the two ground states, i.e., $\sim 0.1\,\text{eV}$.

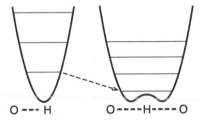

**Fig. 7.4** The potential and the vibrational states of a proton bound to oxygen (*left*). A proton between two oxygen atoms sees a broader potential (*right*); the vibrational ground state of the composite system is energetically a bit lower than in the left potential

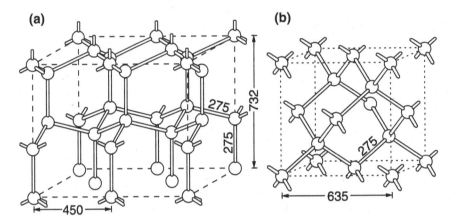

**Fig. 7.5** The structure of ice crystal. The *circles* are oxygen atoms, the *long connecting lines*, measured in pm $= 10^{-12}$ m, correspond to hydrogen bridge bonds. The hydrogen atoms oscillate or tunnel between two oxygen atoms. The crystal is shown in two projections to bring out the empty intermediate spaces

## 7.2.4   Ice

The hydrogen bridge bond leads to a wide variety of crystal structures in ice. Ice in the region of 0 °C has a very loose structure (Fig. 7.5) because, in this state, each oxygen atom has only four hydrogen bridge bonds to its neighbours. This is why there are empty spaces in the rings that form a hexagonal lattice. This explains why ice is lighter than water.

## 7.2.5   Specific Heat

In melting, although the crystal falls apart, clusters of water molecules remain because of the hydrogen bridge bonds. In liquid water, an oxygen atom can be, for a time, bound to up to five neighbours. From melting until evaporation, the clusters become ever smaller and fewer. The main part of the specific heats is needed to break up the hydrogen bridge bonds. The specific heat per water molecule is $9k_B$, while the typical value for liquids and solids is $3k_B$. The latent heat of fusion, heating to boiling point and the latent heat of evaporation together amount to $54.5$ kJ/mol $= 0.6$ eV/molecule. This number equals, on average, two bonds per oxygen atom ($0.3$ eV/bond) – in surprisingly good agreement with our rough estimate.

## 7.3 Hydrogen Bridge Bond in Biology

The important biological processes in the cell are controlled by DNA molecules and proteins. Here, various specific interactions between the various molecules take place. The properties of these are not just fixed by the chemical structure of the molecules but primarily by a well defined three-dimensional structure. The large variety of molecular architectures are first and foremost made possible through hydrogen bridge bonds.

The structure of the proteins may be divided according to their complexity into four principal categories: primary, secondary and tertiary structures as well as higher levels.

### 7.3.1 Primary Structures

Amino acids are attached to each other by peptide bonds and so form a polypeptide chain. The peptide bond is a covalent C–N bond (Fig. 7.6).

The polypeptide chain may be rotated around the covalent bond axis between the nitrogen and carbon atoms – defined by the angle $\Phi$ – and between two carbon atoms ($C_\alpha$–$C'$) – defined by the angle $\Psi$ (Fig. 7.6). The sequence of the amino acids is also called the primary structure.

### 7.3.2 Secondary Structure

The secondary structures are at a higher organisational level. Their elements are spatially ordered structures of the main chain, which only accept well-defined values of the angles $\Phi$ and $\Psi$. One distinguishes between different secondary structure elements. In proteins, the $\alpha$ helix is the most common, but the $\beta$-pleated sheet is also often encountered.

**Fig. 7.6** Schematic
representation of the degrees
of freedom in a polypeptide
chain. The labeling of the
carbon atoms as $C'$ and $C_\alpha$
corresponds to their place in
the chain

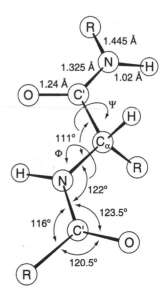

**Fig. 7.7** The hydrogen
bridge bond connects an
amino acid to its fourth
nearest neighbour. This is
responsible for the geometry
of the $\alpha$-helix. The *dark
spheres* are carbon atoms,
the *pale spheres* nitrogen
atoms and the *small spheres*
with springs on them are
hydrogen atoms in a
hydrogen bridge bond

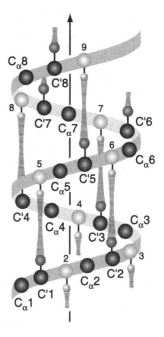

### 7.3.3 α Helix

The production of α-helical structures may be understood as a phase transition between an unfolded random coil state and the helical state. It is here assumed that a core of four neighbouring amino acids first cooperatively enters the helix state through creation of hydrogen bridge bonds, and then, through further hydrogen bridge bonds, completes the full helix (Fig. 7.7).

### 7.3.4 β-Pleated Sheet

This structure is also primarily stabilised by the hydrogen bridge bond. The primary difference from the α helix is that, in the β-pleated sheet, the interactions are between amino acids, which are far apart along the polymer chain (Fig. 7.8).

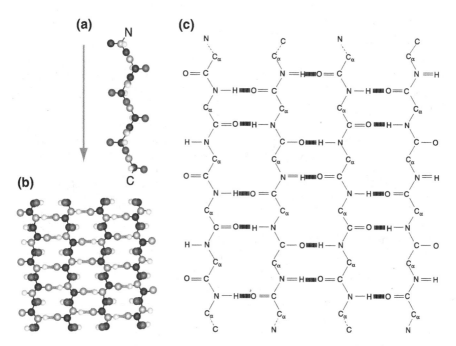

**Fig. 7.8** The neighbouring amino acids in the three-dimensional structure of the β-pleated sheet are widely separated along the polypeptide chain. (a) A segment of the polypeptide chain, (b) several neighbouring segments, (c) the corresponding chemical formula

**Fig. 7.9** Sketch of the three-dimensional structure of an enzyme (triphosphatisomerase, higher level) that is symmetrically constructed from four tertiary structures

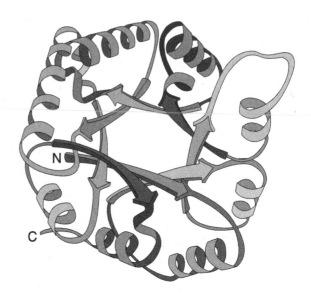

## 7.3.5   Tertiary Structure and Higher Levels

The tertiary structures of proteins are three-dimensional structures that are built up from secondary-structure elements. These protein blocks are usually responsible for a specific biological function.

Globular proteins are constructed from several tertiary structures and can perform various biological functions. Figure 7.9 shows the three-dimensional structure of an enzyme. One can clearly see how this protein is made from four identical molecules of the tertiary structure.

If Occam's statement holds anywhere, then surely for the multiplicity of biological structures.

In summary, we can say that amino acids are the building blocks of proteins, which are bound by covalent bonds into polypeptide chains. Hydrogen bridge bonds produce the links between the structures in the chain and thus enables the huge variety of specific three-dimensional structures in the proteins. This bond is particularly suited to it, as its relative weakness enables these structures to be rapidly built up and rearranged.

## Literature

T.E. Creighton, *Proteins* (Freeman, New York, 1993)

W. Hoppe et al., *Biophysics* (Springer, Berlin, 1983)

G.E. Schulz, R.H. Schirmer, *Principles of Protein Structure* (Springer, Berlin, 1985)

# Chapter 8
# Cold Neutron – Spectroscopy of the Solid State

*Die Wahrheit ist konkret.*

Bertolt Brecht

Cold neutron scattering is the method par excellence for the investigation of condensed matter excitations in both the solid and the liquid phases. The neutrons primarily interact with atomic nuclei, and thus, the form of the excitation mechanism is well defined. Measurements of the energies of the incoming and scattered neutrons together with the scattering angle fully determine the kinematics of the inelastic scattering. The momentum, $\mathbf{q}$, transferred to the system is given by

$$\mathbf{q} = \mathbf{p} - \mathbf{p}', \qquad (8.1)$$

and the energy transfer is

$$\hbar\omega_q = \frac{p^2}{2M_{\mathrm{n}}} - \frac{p'^2}{2M_{\mathrm{n}}}. \qquad (8.2)$$

Here, we have assumed that the system under investigation is at a sufficiently low temperature and that the neutrons cannot pick up any energy from the system. The dependence of the excitation energy, $\hbar\omega_q$, on the momentum transfer, $q$, is called the dispersion relation.

High flux reactors with a deuterium cooled core are the most used source of neutrons. Because the neutrons are not quite cooled to the temperature of the fluid deuterium, their spectrum corresponds to a Maxwell-like distribution, with an energy peak corresponding to roughly 40 K. In precision experiments, one measures the beam energy and the energy of the scattered neutrons with the help of Bragg scattering off the crystal. In Fig. 8.1, we sketch the dispersion curves for an ideal, isotropic crystal, for glass, for a Fermi liquid (liquid $^3$He) and for a superfluid Bose liquid

© Springer-Verlag GmbH Germany 2017
B. Povh and M. Rosina, *Scattering and Structures*,
Graduate Texts in Physics, DOI 10.1007/978-3-662-54515-7_8

(superfluid $^4$He). In each of these cases, the dispersion curves at small $q$ correspond to phonon excitations. For large phonon wavelengths, the dispersion relation is well described by

$$\hbar\omega_q = vq, \tag{8.3}$$

where $v$ is the phonon speed. At short wavelengths, comparable with the interatomic separation $a$, additional excitation modes appear. The phonon picture of the excitation only makes sense as long as the phonon wavelength satisfies $\lambda \geq 2a$. Because these separations are comparable in both the liquid and solid states, it is useful to give the momentum in units corresponding to a phonon, with $\lambda = 2a$. This unit is well defined for crystals and is given by $[\hbar\pi/a]$; for liquids, we assume $a = \sqrt[3]{m_{\text{Atom}}/\rho}$. The momentum dependence of the energy of long wavelength phonons is given by (8.3). If we quote the phonon energy in units of $[v\hbar\pi/a]$, then we largely cancel the dependence on the material's properties in the comparison of the acoustic phonon branch. The unit chosen in this fashion lies between 1.5 meV for liquid helium and 10 meV for metals. Both the similarities and the differences between the dispersion curves in Fig. 8.1 are immediately visible.

In this chapter, we will only treat the scattering of cold neutrons off crystals and glasses; we will discuss the scattering off quantum liquids in Chap. 10.

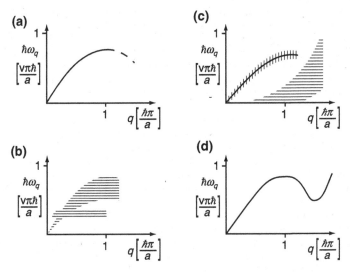

**Fig. 8.1** Four typical dispersion curves for (**a**) a crystal, (**b**) a glass, (**c**) a Fermi liquid (normal liquid $^3$He) and (**d**) a superfluid Bose liquid (superfluid $^4$He). In all four cases, the unit is that of a momentum corresponding to the wavelength $\lambda = 2a$, where $a$ is the average interatomic separation. The energy scale is given in units $[v\hbar\pi/a]$

## 8.1 Dispersion Relations for Crystals

The most researched and best understood dispersion relations are those of crystals, although we cannot present a general derivation of them here. There is only space *on the back of an envelope* for the simplest possible case of a cubic crystal composed of identical atoms with a single, regular separation $a$.

It should be noted that the momentum, $\mathbf{q}$, is completely transferred to the crystal. The internal excitation of the lattice – phonons – are relative movements of the atoms and carry no momentum. Nonetheless, one can assign them a pseudo-momentum, which is conserved modulo $2\pi\hbar/a$ for each component of the pseudo-momentum, $q_i^{\text{pseudo}}$, in a cubic lattice. The phonon wavelength is bounded by the lattice constant $a$, so the pseudo-momentum, $q^{\text{pseudo}}$, and the transferred momentum, $q$, are related by

$$q_i^{\text{pseudo}} = q_i - n_i \frac{2\pi\hbar}{a} . \tag{8.4}$$

Because the internal excitations of the crystal are always described by the dispersion relation for $q^{\text{pseudo}} \leq \pi\hbar/a$, we will simply denote the pseudo-momentum by $q$.

The dispersion relation depends on the phonon propagation direction. In a cubic crystal, one can reduce the complicated problem of finding a general solution of the equation of motion of phonons to a one-dimensional problem by solely considering propagation in the [100], [110] and [111] directions. In all three cases, the crystal planes move as a whole, albeit with different spring constants and a different separation $a$ between the planes. The equation of motion can be written as follows:

$$M \frac{d^2 u_s}{dt^2} = \sum_j G_{sj}(u_{s+j} - u_s) . \tag{8.5}$$

Here, $G_{sj}$ is the spring constant between plane $s$ and plane $j$, where the index $j$ labels the planes and runs from $-\infty$ to $+\infty$. The displacement $u_s$ can be in the propagation direction – longitudinal polarisation – or in the two orthogonal directions – transverse polarisation. For the propagation direction [100], $G_{si}$ is the same for all three polarisations. The dispersion relations for the longitudinal and both transverse polarisations are identical.

Consider the propagation of longitudinally polarised phonons in the [100] direction and simultaneously assume that the interactions are nonzero only between neighbouring planes. We are looking for a solution to (8.5) of the form

$$u_s(t) = U_q e^{(-i\omega_q t + iqas/\hbar)} . \tag{8.6}$$

Substituting into (8.5) leads to a relation between $\omega_q$ and $q$,

$$\omega_q^2 = \frac{2G}{M}(1 - \cos qa/\hbar) . \tag{8.7}$$

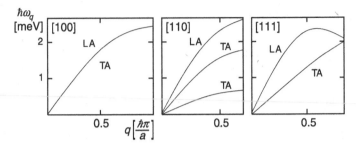

**Fig. 8.2** Dispersion curves for a sodium crystal at 90 K in the [100], [110] and [111] directions. Here, LA denotes the longitudinal and TA the transversal phonon polarisations. The energy is in meV and the momentum transfer is in units of $[\hbar\pi/a]$

As previously mentioned, the dispersion relation in the [100] direction is the same for longitudinally and transversely polarised phonons. For other directions, this is not the case. We will now illustrate the dispersion curves obtained by means of neutron scattering through the example of a monoatomic sodium crystal.

### 8.1.1   Sodium Crystal

The crystal structure of sodium at room temperature is cubic body centred and so the results of the previous section may be applied. Figure 8.2 displays the dispersion curves for the selected propagation directions [100], [110] and [111].

The dispersion curves depend on the propagation directions and the phonon polarisation. Clearly, the spring constants generally differ for the different directions and phonon polarisations. Furthermore, the separations between the planes depend on the propagation direction. If, however, we represent the dispersion curves in units of $[\hbar\pi/a]$ for the momentum and $[v\hbar\pi/a]$ for the energy, then there is no longer any significant dependence on the propagation direction. Our sketch (Fig. 8.1a) shows the universal dispersion curve for crystals.

### 8.1.2   Potassium Bromide Crystal

In crystals with various sorts of atoms, e.g., in alkali halides, there are, in addition to the acoustical phonon branches just described, also optical branches, which correspond to the excitation modes in which neighbouring atoms move with the opposite phase. These excitations also possess the wavelengths between $\lambda = \infty$ and $\lambda = 2a$,

**Fig. 8.3** Dispersion curves for a potassium bromide crystal in the [111] direction. LO and TO denote the longitudinal and transversal polarisations, respectively, of the optical branch, while LA and TA signify their acoustic branch counterparts. The energy scale is in meV and the momentum transfer is in units of $[\hbar\pi/a]$

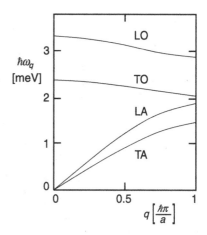

corresponding to $q = 0$ and $q = \hbar\pi/a$, respectively (Fig. 8.3). These optical excitations exist in all crystals with more than one atom in the primitive unit cell. They may be compared with the giant dipole resonance in nuclei (cf. Sect. 14.3)!

## 8.2 Localised Vibrational Mode

Consider an isolated crystal defect with an atom that is either lighter or heavier than all the other atoms but has exactly the same spring constant. In analogy to (8.5), the equation of motion is then

$$(M + \delta M \delta_{s,0})\frac{d^2 u_s}{dt^2} = \sum_j G_{sj}(u_{s+j} - u_s). \tag{8.8}$$

For simplicity, we will here too describe the system as a one-dimensional chain. The generalisation to three dimensions is obvious: the displacements, $u$, and pseudo-momenta, $q$, become vectors and the equations are otherwise unaltered.

We again expand the displacements, $u_s$, in unperturbed eigenmodes – phonon fields $U_q$,

$$u_s(t) = \sum_q U_q e^{(-i\omega t + iqas)}, \tag{8.9}$$

and find the secular equation

$$\omega^2 \begin{pmatrix} M + \delta M/N & \delta M/N & \delta M/N & \cdots \\ \delta M/N & M + \delta M/N & \delta M/N & \cdots \\ \delta M/N & \delta M/N & M + \delta M/N & \cdots \\ \vdots & \vdots & \vdots & \ddots \end{pmatrix} \begin{pmatrix} U_1 \\ U_2 \\ U_3 \\ \vdots \end{pmatrix}$$

$$= \begin{pmatrix} M\omega_1^2 \, U_1 \\ M\omega_2^2 \, U_2 \\ M\omega_3^2 \, U_3 \\ \vdots \end{pmatrix}.$$

(8.10)

The nondiagonal matrix elements, $\delta M/N$ come from the Fourier transformation of the localised mass terms, $\delta M \delta_{s,0}$.

To solve (8.10), we express each coefficient as a sum over all the other coefficients:

$$U_q M (\omega^2 - \omega_q^2) = -\frac{\delta M}{N} \omega^2 \sum_p U_p ,$$

(8.11)

where $\sum_p U_p$ is a constant. We sum both sides over all $N$ coefficients, taking into account that $\sum_q U_q = \sum_p U_p$, and dividing by this sum obtain the relation

$$1 = -\frac{\delta M/N}{M} \sum_q \frac{\omega^2}{\omega^2 - \omega_q^2} .$$

(8.12)

The solutions of this equation may best be graphically represented (Fig. 8.4). The right-hand side of (8.12) has poles at the points $\omega = \omega_q$. The solutions $\omega_q'$ are found

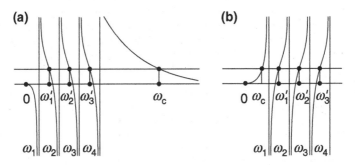

**Fig. 8.4** Graphical representation of the solution of the secular equation (8.10): (**a**) for a solute atom with a smaller mass ($\delta M < 0$), the collective state is promoted from the acoustic phonon branch to higher energies; (**b**) for a solute atom with a larger mass ($\delta M > 0$), the collective state appears at the lower edge of the acoustic branch and does not correspond to a localised excitation

where the right-hand side is unity. The new eigenfrequencies are marked on the abscissa. The $(N - 1)$ eigenvalues are trapped between the unperturbed frequencies, $\omega_q$. The outlier, denoted by $\omega_C$, is the collective state. By collective, we mean that it is a superposition of many unperturbed phonon states.

To describe the collective state, we employ the same formalism as for the pion (Chap. 6) and the giant resonances (Chap. 14) to bring out the analogies. In this chapter, frequencies rather than energies occur, but $E = \hbar\omega$ holds. The frequencies appear quadratically in the secular equation because the equation of motion in this chapter is a differential equation, which is second order in time; in Chaps. 6 and 14, on the other hand, the Schrödinger equation is used, which is first order in time, and the energies appear linearly.

For a smaller mass of the solute atom ($\delta M < 0$), the collective state lies above the acoustic phonon band and therefore cannot behave as a propagating wave (we will not prove this theorem here). This state corresponds to a localised, standing wave (Fig. 8.5).

For a heavier solute atom ($\delta M > 0$), the collective state lies at the lower edge of the acoustic phonon band and is not localised!

The secular equation (8.10) was derived for a monoatomic crystal with an acoustic phonon band. Di- and polyatomic crystals have an additional optical phonon band. A secular equation can also be obtained for this case. For a heavy solute atom, the collective state is lowered, as in the case of the pion (Chap. 12). The localised mode is then found below the optical band and can also appear inside the acoustic band – as a resonance embedded in the phonon continuum (Fig. 8.6).

The eigenfrequency of a localised or resonant impurity mode may be directly observed through optical absorption in the infrared regime.

**Fig. 8.5** Localised mode of a light solute atom above the phonon band

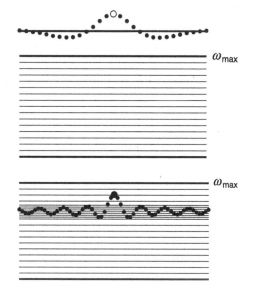

**Fig. 8.6** Resonant mode of a heavy solute atom that passes from the optical phonon branch into the phonon band

There is nothing exotic about these localised excitation modes in crystals; not just solute atoms and dislocations but also interfaces in crystals can produce localised excitations.

## 8.3  Dispersion Relations for Amorphous Substances

We will only qualitatively discuss the dispersion relations of amorphous substances. Because there are still no generally accepted standard dispersion curves for amorphous substances in the literature, we will restrict ourselves to the sketch in Fig. 8.1b, which we will now briefly discuss. For long-wavelength phonons, the disorder at the interatomic scale is not noticeable and one might expect that, at small momentum transfer, the dispersion curve would resemble that of crystals. This is, however, not the case. At small energy transfers, not just phonons but additional excitations are significant. An atom at a lattice site of the crystal is in a harmonic oscillator potential. In an amorphous substance, on the other hand, the potential around the atom is irregular. Generally, the potential has two or more minima. The atoms tunnel from one minimum to the other. This mechanism produces low-energy excitations that coexist with the long-wave phonons. For slightly higher excitations, $\hbar\omega \approx 1 - 2\,\text{meV}$, when the atom is in a wide potential (Fig. 8.7), the excitations accumulate. This accumulation of excitations in the energy range of 1–2 meV can be very clearly seen in inelastic neutron scattering. The peak visible in the measured spectrum is called the bosonic peak. In Fig. 8.1b, we recognise the bosonic peak in the dispersion curve in a narrow energy range with a wide momentum transfer.

At still higher excitations, we can view the disorder at the interatomic scale as localised imperfections, and the smearing of the dispersion curves is due to localised vibrational modes.

The dispersion curves for amorphous substances do not display such simple and attractive properties as those of crystals. Therefore, one cannot learn much from them and it is not surprising that they are not presented.

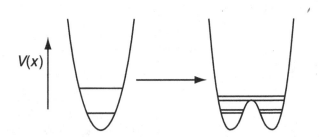

**Fig. 8.7** Interatomic harmonic oscillator potential in a crystal (*left*) and in an amorphic substance (*right*). Due to the disorder in amorphic media, an atom is not well localised – it can tunnel between the potential minima

## 8.4 Specific Heat

### 8.4.1 Crystalline Substances

Specific heat is defined by

$$C_V = \left(\frac{\partial U(T)}{\partial T}\right)_V , \tag{8.13}$$

where $U(T)$ is the internal energy, i.e., the total energy of the phonons of the solid body at temperature $T$. Denoting the phonon state density by $\mathcal{D}(\omega)$ and recalling that phonons obey Bose–Einstein statistics, the expression for the internal energy is

$$U(T) = \int_0^{\omega_D} \hbar\omega \mathcal{D}(\omega) \frac{d\omega}{e^{\hbar\omega/k_B T} - 1} . \tag{8.14}$$

We want to calculate the phonon state density and the cut-off parameter $\omega_D$ in the Debye approximation. We use a linear relation, $\hbar\omega = vq$, where $v$ is the speed of sound, in the dispersion relation. Strictly speaking, this relation only holds for large wavelengths. Let us first calculate the state density for the individual phonon branches,

$$\mathcal{D}(\omega)d\omega = \frac{V 4\pi q^2 dq}{(2\pi\hbar)^3} = \frac{V}{2\pi^2} \frac{\omega^2}{v^3} d\omega . \tag{8.15}$$

Different phonon branches, the longitudinal and the two transverse ones, have different speeds of sound. A simple way to take these speeds of sound for the different phonon branches into account in (8.15) is to introduce an averaged Debye speed, $v_D$,

$$\frac{3}{v_D^3} = \frac{1}{v_l^3} + \frac{2}{v_t^3} . \tag{8.16}$$

The cut-off frequency – the Debye frequency – $\omega_D$, depends on the spring constants, the masses of the atoms and the lattice constants and consequently varies from crystal to crystal. Furthermore, $\omega_D$ also depends on the polarisation of the phonon. In the Debye approximation, all of this dependence is replaced by a single cut-off parameter. In this approximation, the internal energy is then

$$U(T) \propto \int_0^{\omega_D} \frac{\hbar\omega^3}{e^{\hbar\omega/k_B T} - 1} d\omega . \tag{8.17}$$

We choose the normalisation such that, as $T \to \infty$, the specific heat takes on the value $C_V = 3R$. Instead of $\omega_D$, we introduce the Debye temperature, $\Theta$, via the relation $\hbar\omega_D = k_B\Theta$ and use the notation $x = \hbar\omega/(k_B T)$ and $x_D = \hbar\omega_D/(k_B T) = \Theta/T$. The specific heat, from (8.13), in the Debye approximation is then

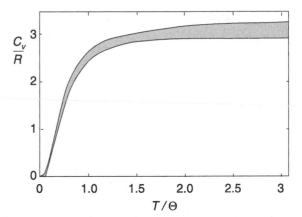

**Fig. 8.8** Molar specific heat of a series of substances (Pb, FeS$_2$, KCl, Fe, Na, CaF$_2$, Zn, NaCl, Ag, Tl, KBr, Ca, Cu, C, Al, Cd) as a function of the reduced temperature $T/\Theta$. All the experimental data lie on a universal curve between the displayed bounds

$$C_V = 9R\left(\frac{T}{\Theta}\right)^3 \int_0^{x_D} \frac{x^4 e^x}{(e^x - 1)^2}dx \,. \tag{8.18}$$

The normalisation follows from the integration at $T \to \infty$, where we have $x \to 0$, and one may expand the exponential function through the integration range,

$$\left(\frac{T}{\Theta}\right)^3 \int_0^{x_D} \frac{x^4 e^x}{(e^x - 1)^2}dx \approx \left(\frac{T}{\Theta}\right)^3 \int_0^{x_D} \frac{x^4}{(1 + x - 1)^2}dx$$
$$= \left(\frac{T}{\Theta}\right)^3 \int_0^{x_D} x^2 dx = \frac{1}{3}\,, \tag{8.19}$$

which is the explicit form of the Debye formula.

Figure 8.8 displays the excellent agreement of the Debye formula with experimental measurements.

### 8.4.2  Amorphous Substances

At high temperatures, the specific heats of amorphous substances – just as with crystals – are described by the Dulong–Petit law because, of course, in both cases, all vibrational degrees of freedom are excited. The deviation from crystals is especially noticeable at low temperatures. The phonon concept works well for amorphous substances too – so long as their wavelengths are much longer than the average separation of the atoms. However, it is just at low temperatures where the behaviour of the specific heats of amorphous substances deviates from the Debye theory. Experimentally, the specific heats at low temperatures are larger than in crystals. Obviously, in amorphous substances, there are additional excitation modes, two of which we have already mentioned: tunnelling modes and bosonic peak modes.

# Literature

C. Kittel, *Introduction to Solid State Physcis* (Wiley, New York, 1995)

J.M. Ziman, *Principles of the Theory of Solids* (Cambridge University Press, Cambridge, 1979)

# Chapter 9
# Quantum Gases – Quantum Degeneration

*Wissenschaft, die nicht vermittelt wird, ist tot.*

Ranga Yogeshwar

The models of quantum gases were already developed in the 1920s: the Fermi gas model for degenerate Fermi systems and the model of Bose condensates for degenerate bosonic systems. Both models may also be reasonably well applied to the description of quantum liquids.

Because of the successful production of quantum gases in a metastable state at temperatures far below the μK regime, it is possible to directly investigate their properties. These experiments have become very fashionable in recent years because, using quantum gases, quantum mechanical effects can be observed in macroscopic systems.

### Production of Cold Gases

A Bose–Einstein condensate is usually produced in several steps. The first step is the cooling and subsequent capture of the atoms with laser light at very low densities. Laser cooling fails, though, at densities where the average separation corresponds to an optical wavelength; light is then no longer absorbed and re-emitted by individual atoms but rather by atomic clusters. One can reach greater densities if one stores the atoms in a magnetic trap. In the final phase of cooling, one lets the more energetic atoms evaporate out of the trap. The remaining lower energetic atoms redistribute their energy through collisions and thus lower their temperature. The phase of the condensate can be nicely experimentally demonstrated by a time-of-flight measurement. The magnetic trap is switched off and the atoms may fly freely for a few milliseconds. Subsequently, one illuminates the atoms with laser light and photographs the shadow of the atomic cloud. From various flight times, one obtains the velocity distribution of the gas. The velocity distribution indeed corresponds to the Bose–Einstein expression.

© Springer-Verlag GmbH Germany 2017
B. Povh and M. Rosina, *Scattering and Structures*,
Graduate Texts in Physics, DOI 10.1007/978-3-662-54515-7_9

**Fig. 9.1** Symbolic
representation of the
occupation of the states of a
Fermi and a Bose gas at
$T = 0\,\mathrm{K}$

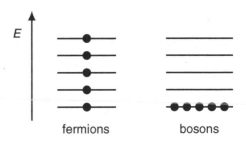

The evaporation cooling method does not work for fermions. One can only store atoms with the same magnetic quantum number in a magnetic trap. The Pauli principle forbids identical fermions simultaneously occupying the same point of phase space and the probability of a collision becomes ever smaller as the temperature decreases. This is why cooling via collisions does not work for a pure Fermi gas in the same way as for gases made of bosons. Such cooling does work, though, if one simultaneously cools two fermionic gases. This was demonstrated for the first time in 1999 by B. DeMarco and D. Jin. In their experiment, atoms with $^{40}$K nuclei were used. The $^{40}$K nucleus with $J^{\pi} = 4^{-}$ and an unpaired $s_{1/2}$-electron couple in the atomic ground state to a total angular momentum, $F^{\pi} = 9/2^{-}$. (The $F = 9/2$ state lies below the $F = 7/2$ state because $^{40}$K has a negative magnetic moment.) The trap is filled with two Fermi gases, one with atoms in the $m_{F} = -9/2$ hyperfine state and the other in $m_{F} = -7/2$. The Pauli principle does not limit different hyperfine states and they can collide like bosons. The two gases cool each other down. In the experiment we are describing, the Fermi gas mixture was cooled to below the Fermi temperature ($T \approx T_{F}/2 \approx 300\,\mathrm{nK}$) and the degeneracy was observed. In Fig. 9.1, the occupation of the states for ideal Fermi and Bose gases is symbolically represented.

The average separations between atoms or molecules in a gas must be small enough, compared with their sizes, and the densities must be sufficiently low that only two particle collisions are possible. Under these conditions, one can produce metastable Fermi gas systems in the laboratory. If the temperature and density of the gas fulfill the requirements for degeneracy, the low-energy states are occupied. The experimental detection of a Fermi gas is significantly harder than for a bosonic gas because it does not have a phase transition and the change is continuous.

## 9.1 Fermi Gas

In what follows, we want to determine the condition for the occurrence of Fermi degeneracy. A rough estimate may be obtained by setting the de Broglie wavelength equal to the average separation of the atoms, $\lambda_{T} \approx d$. In thermal systems, one defines a thermal de Broglie wavelength, $\lambda_{T} = 2\pi\hbar/p$, corresponding to the momentum, $p = m\sqrt{\langle v^{2}\rangle} = \sqrt{2\pi m k T}$.

Because of the Pauli principle, the fermions do not overlap in space. In a single magnetic substate of a fermionic gas, we expect the transition to a degenerate state to occur when

$$V \approx N\lambda_T^3 = N \frac{(2\pi\hbar)^3}{(2\pi mkT)^{3/2}}.$$ (9.1)

The relation between the transition temperature and the particle density may be rewritten as

$$kT \approx \frac{2\pi\hbar^2}{m} \left(\frac{N}{V}\right)^{2/3}.$$ (9.2)

### 9.1.1 Fermi Energy, Fermi Momentum, Fermi Temperature

The scale used in a degenerate fermion system is the Fermi energy, $E_F$, or the related Fermi momentum, $p_F$, or the Fermi temperature, $T_F$.

In a Fermi gas at $T = 0$, all the states below the Fermi energy, $E_F = p_F^2/(2m)$, are occupied. In a volume $V$, the number of fermions below the Fermi momentum, $p_F$, for nonrelativistic particles is then

$$N = \kappa \frac{4\pi}{3} \frac{p_F^3 V}{(2\pi\hbar)^3},$$ (9.3)

where $\kappa$ is the number of magnetic substates available in the Fermi gas. This implies

$$p_F = (6\pi^2\hbar^3)^{1/3} \left(\frac{N}{\kappa V}\right)^{1/3}$$ (9.4)

and

$$E_F = kT_F = \frac{1}{2m} (6\pi^2\hbar^3)^{2/3} \left(\frac{N}{\kappa V}\right)^{2/3}.$$ (9.5)

As one can easily show by integrating over all the Fermi states below $E_F$, the average kinetic energy is

$$\langle E \rangle = \frac{3}{5} E_F.$$ (9.6)

### 9.1.2 Transition to a Degenerate Fermi Gas

The transition from a normal to a degenerate gas takes place when the atoms start to overlap. This is the case when the de Broglie wavelength roughly corresponds to the average separation between the atoms. A somewhat more precise estimation now follows.

Defining the average separation $d$ by

$$\left(\frac{N}{\kappa V}\right) = \frac{1}{2d^3} , \tag{9.7}$$

we obtain, for particles with spin $s = 1/2$ and $\kappa = 2$, the following relation between the average kinetic energy and the average separation in the degenerate state:

$$\langle E \rangle = \frac{3}{5}(3\pi^2)^{2/3}\frac{\hbar^2}{2md^2} . \tag{9.8}$$

This implies that $d = 1.49\lambda$, where $\lambda$ is the de Broglie wavelength, which corresponds to the average kinetic energy of the particles.

As previously mentioned, the transition to a degenerate Fermi gas does not take place via a sharp phase transition because its speed depends on the cooling method. However, the degeneracy of a Fermi gas has been experimentally demonstrated.

## 9.2  Bosonic Gas

The degeneracy of a bosonic gas occurs – as with a Fermi gas – when the de Broglie wavelength is comparable with the average separation of the atoms, $d \approx \lambda_T$. In contrast with a Fermi gas, for a bosonic gas, there is a phase transition between the normal gas phase and the condensate. This transition is theoretically particularly easy to describe and may be used as a model for complicated cases in solid state physics and also for phase transitions, such as chiral symmetry breaking or the Higgs model. We will therefore briefly describe this transition, though only for an ideal gas in a large volume. Experiments are carried out in traps, where the atoms are held together by a confinement potential. The description is a bit different, but the physics remains the same.

### 9.2.1  Bose–Einstein Condensation

The occupation of the states in an ideal bosonic gas is given by the distribution function

$$N_\varepsilon = \frac{1}{e^{(\varepsilon-\mu)/(kT)} - 1} . \tag{9.9}$$

Here, $\varepsilon$ denotes the energy of the state and $\mu$ is the so-called chemical potential. The latter takes the energy of the system into account, which depends on the temperature and the particle number, and is defined by

$$\mu = \left(\frac{dE}{dN}\right)_{V,S=\text{const}}.$$ (9.10)

The distribution function must be positive, $N_\varepsilon \geq 0$, and thus $\mu \leq \varepsilon_0$. For an ideal gas, the energy of the ground state, $\varepsilon_0 = 0$, and consequently $\mu \leq 0$. The total number of bosons in the gas is

$$N = N_0 + \int_0^\infty f(\varepsilon) N_\varepsilon d\varepsilon.$$ (9.11)

$N_0$ is the number of particles in the ground state with energy $\varepsilon_0 = 0$, and $f(\varepsilon)$ is the available phase space. The spatial extent in the experiment is given by the confinement potential. Here we want, though, to give the phase space just for a free gas,

$$f(\varepsilon) d\varepsilon = \frac{4\pi p^2 dp V}{(2\pi\hbar)^3}.$$ (9.12)

Because $dp/d\varepsilon = m/p$, we have

$$f(\varepsilon) = \frac{1}{(2\pi)^2} \left(\frac{2m}{\hbar^2}\right)^{3/2} V\sqrt{\varepsilon}.$$ (9.13)

For the case of a confining potential, the expression for the phase space is essentially only altered in the exponent of the energy dependence. Let us now consider the phase transition from a normal gas to a condensate. The phase transition takes place when, each time a particle is added, it enters the ground state. Then the energy of the system in the case of an ideal gas ($\varepsilon_0 = 0$) is not altered! The temperature at which $\mu = 0$ is the critical temperature, $T_c$. In Fig. 9.2, the dependence of the chemical potential on the temperature is sketched. For temperatures $T \leq T_c$, many

**Fig. 9.2** The dependence of the chemical potential, $\mu$, on the temperature. The temperature enters both the ordinate, $\mu/kT$, and the abscissa, $d/\lambda_T \propto \sqrt{T}$. The average separation of the particles is $d$, the thermal Compton wavelength is $\lambda_T$

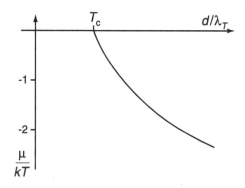

particles are accepted into the ground state and the number of particles that are not in the ground state can be easily calculated as

$$N - N_0|_{\mu=0} = \int_0^\infty f(\varepsilon) \frac{d\varepsilon}{e^{\varepsilon/(kT)} - 1}$$

$$= \frac{1}{(2\pi)^2} V \left(\frac{2mkT}{\hbar^2}\right)^{3/2} \int_0^\infty \frac{\sqrt{x}\, dx}{e^x - 1} .$$

(9.14)

The value of the integral is 2.612. The critical temperature is determined from the limit $N_0 \rightarrow 0$, and thus depends on the density, $T_c \propto (N/V)^{2/3}$. The probability of finding a particle in the ground state is roughly sketched in Fig. 9.3 and is given by

$$\frac{N_0}{N} = 1 - \left(\frac{T}{T_c}\right)^{3/2} .$$

(9.15)

Equation (9.15) is only correct for a free gas. Experiments are, however, carried out in a trap. In a confinement potential, the alteration to (9.15) is only in the exponent, i.e.,

$$\frac{N_0}{N} = 1 - \left(\frac{T}{T_c}\right)^3 .$$

(9.16)

The phase transition between a normal gas and the condensate can be very easily demonstrated mathematically. In more complicated physical systems, in which the phase transition cannot be so directly demonstrated, the mathematical treatment still follows the same pattern. In Fig. 9.4, the occupation of the levels of a bosonic gas in three temperature domains is sketched. Below $T_c$, the ground state is occupied by many atoms. The occupation number of the atoms in the ground state serves as an order parameter of the condensed phase.

**Fig. 9.3** The temperature dependence of the probability of finding a boson in the ground state

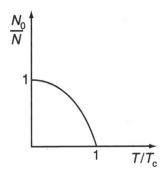

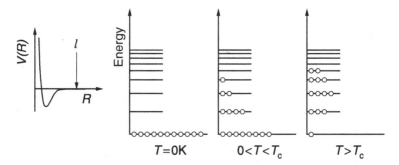

**Fig. 9.4** An illustration of the occupation of the energy levels of an ideal bosonic gas at $T = 0$, $0 < T < T_c$ and $T > T_c$. It is symbolically indicated that the average separation, $l$, of the atoms is much larger than the range of the interatomic potential

A gas of noninteracting bosons enters a Bose–Einstein condensate at a finite temperature! This is because arbitrarily many bosons may occupy the ground state. This statement is, though, not true for systems in fewer than three dimensions. Three-dimensional phase space (9.12) is needed for bosonic gases to have a phase transition at finite temperature.

Bosonic systems differ from our paradigm example, the ferromagnets. Without positive feedback, the phase transition from a paramagnetic to a ferromagnetic state would first take place in the limit $T \to 0$.

## 9.3  Coherent Photon Gas – Laser

Besides its great success as a research and technological tool, laser light offers also many conceptual insights.

Here we concentrate on the aspect of coherence with the consequence of a very sharp spectral line and very sharp solid angle of propagation. Although all photons are in the same quantum state, this is not the case of Bose-Einstein condensation – the density of photons is not that high – but it is rather the case of coherent phases of photons. We describe the many-photon wave function in Fock space, as a superposition of zero-photon, one-photon....many-photon states

$$\Phi = (1 + c_1 e^{i\varphi_1} \phi^1 + c_2 e^{i\varphi_2} \phi^2 + c_3 e^{i\varphi_3} \phi^3 + \cdots) . \tag{9.17}$$

Any device capable of measuring the field, including its phase, must be capable of altering the number of quanta by an indeterminate amount (see Peierls 1979). This fact is theoretically implemented in the form of electromagnetic operators $\hat{\mathcal{E}}$ and $\hat{B}$ which either create or annihilate a photon. For example

$$\mathcal{E} = \langle \Phi | \hat{\mathcal{E}} | \Phi \rangle = \sum_0^\infty \left( c_j^* c_{j+1} e^{-i(\varphi_{j+1}-\varphi_j)} \sqrt{j+1} \langle 0 | \hat{\mathcal{E}} | \phi \rangle + c_{j+1}^* c_j e^{i(\varphi_{j+1}-\varphi_j)} \sqrt{j+1} \langle \phi | \hat{\mathcal{E}} | 0 \rangle \right).$$

$$(9.18)$$

There exists an uncertainty relation for the number of photons $n$ and the phase $\varphi$ of the system: $\Delta n \times \Delta \varphi \sim \frac{1}{2}$, analogous to the uncertainty relation between the angular momentum and orientation angle (This simple relation is of course valid only for small $\varphi$ since $\varphi$ is determined only modulo $2\pi$.)

The uncertainty is smallest for a *coherent state* of photons in which all phase differences are the same, $\varphi_j = j \varphi$.. For monochromatic photons with a frequency $\omega$ this gives a sinusoidal oscillation of the electric (and magnetic) field, similar as in a classical description:

$$\mathcal{E} = \mathcal{E}_0 \cos(\varphi - \omega t). \qquad (9.19)$$

Laser can be described in three steps:

1. *Powering a laser with optical pumping.* A typical scheme is as follows. An atomic, molecular or solid state three-level system is chosen such that an inverse population of the middle level with respect to the ground level can be reached. Energy is pumped into the system by lifting electrons from the ground level to the upper level by means of absorption of light with appropriate frequency $(E_3 - E_1)/\hbar$. Electrons de-excite rapidly by spontaneous photon emission to the middle, metastable level which reaches a high population. Then the middle level is de-excited by stimulated emission to the ground level feeding the laser device with photons of energy $E_2 - E_1$. Because of the inverse population there is more stimulated emission than absorption back from the ground to the middle level. Due to the stimulated emission, all successive photons contribute to the radiation field in phase (Fig. 9.5). The use of helium to pump electrons into a metastable state of neon in the helium-neon laser is an example of such a mechanism.

**Fig. 9.5** The three-level scheme of the optical pumping and stimulated emission

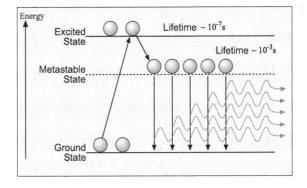

**Fig. 9.6** Coherent photon gas in the cavity with two mirrors

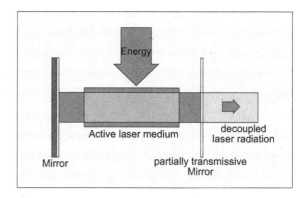

2. The use of a *resonator* selects a chosen frequency and direction of light. Only photons corresponding to the chosen frequency reach a high population and stimulate further emission from the optically pumped material which is usually contained in the resonator itself. (Sometimes two or more frequencies appear.) The resonator consists essentially of a cavity with two mirrors. (Fig. 9.6).

Comparing light emission with sound emission, laser resembles a clarinet. A clarinet also requires energy input, through the mouthpiece and the reed. The feedback of the resonator – the tube – stimulates the vibration of the reed at the frequency of the resonator. There is a difference, however, the clarinet contains only a quarter or few wavelength of sound while the laser resonator contains very many wavelength of light. Therefore the wind instruments in general do not have such a sharp frequency and direction.

3. For application, the EM resonator must be coupled to the user, for example by taking a semitransparent mirror.

# Literature

C.N. Cohen-Tannoudji, Nobel lecture: manipulating atoms with photons. Rev. Mod. Phys. **70**, 707–719 (3 July 1998)

F. Dalfovo et al., Theory of Bose-Einstein condensation in trapped gases. Rev. Mod. Phys. **71**, 463–512 (3 April 1999)

B. DeMarco, D.S. Jin, Onset of Fermi degeneracy in a trapped atomic gas. Science **285**(5434), 1703–1706 (1999)

D.S. Durfee, W. Ketterle, Experimental studies of Bose-Einstein condensation. Opt. Express **2**(8), 299–313 (1998)

L.D. Landau, Lifshits, *Statistical Physics 1* (Pergamon, Oxford)

R.E. Peierls, *Surprises in Theoretical Physics*, Princeton Series in Physics (Princeton University Press, Princeton, 1979)

# Chapter 10
# Quantum Liquids – Superfluidity

$$\pi\acute{\alpha}\nu\tau\alpha\ \rho\epsilon\tilde{\iota}$$

Heraclites

The systems, which may be well described as Fermi liquids, are liquid $^3$He, electrons in metals, nuclei, white dwarfs, neutron stars and, perhaps, quark stars, too. As an example of a bosonic liquid, we will, of course, consider $^4$He. Systems of fermions coupled to boson quantum numbers – Cooper pairs – are also of interest. These are produced, for example, by atom pairing in liquid $^3$He at low temperatures, by electron pairing in metals and by nucleon pairing in nuclei. In this chapter, we will only consider the classic examples of quantum liquids, $^3$He and $^4$He. The rest will be discussed in later chapters.

## 10.1   Normal Liquid $^3$He

The difference between a Fermi gas and a Fermi liquid is demonstrated in a simplified form in Fig. 10.1.

For an ideal Fermi gas – noninteracting atoms – at temperature $T = 0$ K, all states below the Fermi energy are occupied and the states above it are empty (Fig. 10.1a). At finite temperatures, $T > 0$, the Fermi surface is smeared and the smearing measures the actual temperature of the system (Fig. 10.1b). If we also describe Fermi liquids by energy states, then even at $T = 0$, because of interatomic forces, there is no sharp cut-off (Fig. 10.1c). At finite temperatures, the Fermi surface is still more smeared (Fig. 10.1d), due to the appearance of thermal excitations.

The phase transition from gas to liquid is pressure dependent and, under standard experimental conditions, takes place around $T = 3.19$ K. The critical point is at $T_k = 3.32$ K and $p_k = 1.16$ bar. One might expect the properties of liquid $^3$He, as

© Springer-Verlag GmbH Germany 2017
B. Povh and M. Rosina, *Scattering and Structures*,
Graduate Texts in Physics, DOI 10.1007/978-3-662-54515-7_10

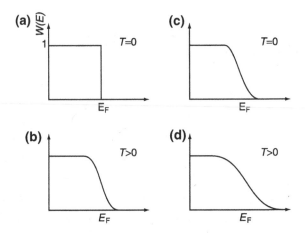

**Fig. 10.1** Occupation of the states of a degenerate Fermi gas at (**a**) $T = 0$, (**b**) $T > 0$ and of a liquid at (**c**) $T = 0$, (**d**) $T > 0$. In both cases, the distributions refer to the states of an ideal gas

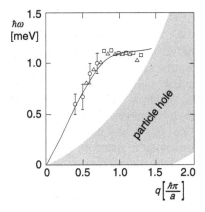

**Fig. 10.2** The scattering of cold neutrons off liquid $^3$He at temperature $T = 120$ mK and saturated vapour pressure shows two dispersion curves. The *lower dispersion curve* corresponds to a particle-hole excitation of $^3$He atoms at the surface of the Fermi sea. The second, which corresponds to phonon-roton excitations, is similar to the dispersion curve in superfluid $^4$He (Fig. 10.5). In contrast with $^4$He, the phonon-roton excitation is strongly damped since it decays into particle-hole excitations. Therefore, we have shaded in this excitation (following Scherm et al.)

a degenerate Fermi system, to be particularly clearly seen in cold neutron scattering. Unfortunately, the cross-section for neutron capture by $^3$He is so large that only semi-quantitative measurements have been made. The dispersion curve, the dependence of the excitation energy on the momentum transfer in liquid $^3$He, Fig. 10.2, distinctly shows two branches. The first corresponds to single-particle excitations, which are better called particle-hole excitations. The relation between the energy loss, $E_{kin}$, to $^3$He and momentum transfer, $p$, for this branch is $E_{kin} = p^2/2M^*$, where $M^*$ is the effective mass of the $^3$He atom. This branch exactly corresponds to neutron scattering off $^3$He at the Fermi surface and it is, apart from $M^*$, identical to the scattering off a Fermi gas at the same temperature. The second branch is an artefact of the liquid state and is analogous to the phonon-roton branch in liquid $^4$He. We will discuss this in the next section.

## 10.2 Superfluid $^4$He

At low temperatures, bosons condense into the lowest or, at least, a very few low-lying states of the system. A condensate is formed when the de Broglie wavelength is larger than the average separation between them (see Chap. 9). Under these conditions, the condensate – even when it has a macroscopic Extension – may be described by a single wave function. This implies that liquids, in particular, will be the first to form condensates when they are cooled. In Fig. 10.3, the dependence of the formation of the condensate on the average separation between the bosons is shown.

For liquid $^4$He, with an average interatomic separation $\approx 0.1$ nm, a condensate forms at a temperature $T \cong 2.17$ K, just below the liquefaction temperature. Even at temperature $T = 0$, a bosonic quantum liquid will not be in a pure bosonic condensate state. Due to the interactions between the atoms, there are, as well as the condensate, also single-particle excitations. In the case of $^4$He at, for example, $T \approx 2$ K, only around 10% of the atoms are in the collective ground state and in collective excitation states. Figure 10.4 schematically shows the occupation of the levels in superfluid helium II at $T = 0$ K, $0$ K $< T < T_\lambda$ and of liquid helium at temperature $T > T_\lambda$. This should be compared with Fig. 9.4 to clarify the differences between the condensates of bosonic gases and bosonic liquids.

The dispersion curve for superfluid $^4$He obtained using cold neutrons is particularly marked (Fig. 10.5) and worth a closer look. A pure phonon excitation would

**Fig. 10.3** The condensate phase appears when the de Broglie wavelength is larger than the average separation of the bosons. A bosonic liquid forms a condensate at significantly higher temperatures than a bosonic gas

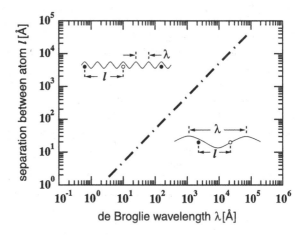

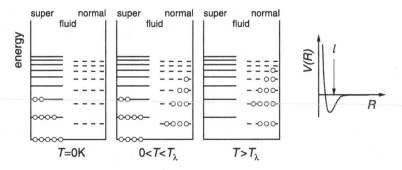

**Fig. 10.4** Sketch of the level occupation in superfluid helium II below $T_\lambda$ and in normal liquid helium above $T_\lambda$. The fact that the average separation, $l$, is comparable with the helium diameter is also symbolically shown

**Fig. 10.5** The dependence of the energy loss on the momentum transferred from neutrons to helium II; the *dashed line* determines the propagation velocity of the rotons

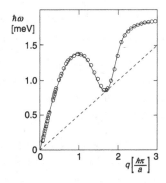

correspond to a monotonically increasing energy-momentum dependence, $E_{ph} = v_{ph} p$. This dependence is shown in Fig. 10.5 as a dashed line. The gradient of the line at $p = 0$ implies a phonon velocity of $v_{ph} \approx 238\,\text{m/s}$.

The deviation from the phonon excitation is attributed to roton excitations. Rotons correspond to quantised vortices in helium, whose formal description is not easy. Therefore, we present here an analogy for rotons that was invented by Feynman.

A passenger wants to get out of a tightly packed street car – a good approximation for the closely packed helium atoms. There are two options to reach the door: either she uses a great deal of force to lift herself up and then reach the door over the heads of her fellow passengers. In a quantum system, this would correspond to an excitation to a higher lying state, which later decays. Alternatively – in a much more economical method – she may ask each of the passengers between her and the door to swap places with her one after another and so slowly reach the door. This second possibility illustrates the idea of quantised vortices.

Rotons cannot be excited below the energy $\Delta_R$. The tangent to the roton curve in Fig. 10.5, $E_R = v_R p$, defines the rotons' propagation velocity, $v_R \approx 58\,\text{m/s}$.

**Fig. 10.6** A sphere with mass $M$ moves with velocity **v** and emits a phonon in direction $\Theta$ with energy $\hbar\omega$ and momentum **p**. For $M \to \infty$, the critical speed is $v_c \approx v_R = 58\,\text{m/s}$

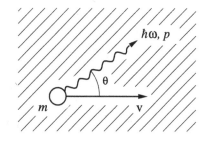

How can superfluidity arise in helium if phonons can pick up energy at arbitrarily small momenta? To explain this, it suffices to show that the viscosity, measured using a sphere at small speeds, is equal to zero.

Consider a sphere of mass $M$, moving in helium ($T = 0$) with speed $v$. Through dissipation, the sphere produces an excitation with energy $\varepsilon$ and momentum $p$, Fig. 10.6.

Energy conservation implies that

$$\frac{1}{2}Mv^2 = \frac{1}{2}Mv'^2 + \varepsilon, \tag{10.1}$$

where $v'$ is the speed after the production of the excitation. Momentum conservation additionally requires

$$M\mathbf{v} - \mathbf{p} = M\mathbf{v}'. \tag{10.2}$$

It is easy to see that the phonon excitation first occurs when the sphere moves with a speed $v \geq v_{\text{ph}}$. Squaring (10.2) implies

$$\frac{1}{2}Mv^2 - \mathbf{v} \cdot \mathbf{p} + \frac{1}{2M}p^2 = \frac{1}{2}Mv'^2 \tag{10.3}$$

or, more briefly,

$$\varepsilon = \mathbf{v} \cdot \mathbf{p} - \frac{1}{2M}p^2. \tag{10.4}$$

The minimal value for the size of the velocity $\mathbf{v}_c$ at which dissipation can take place is when the momentum of the excitation and the velocity of the sphere are parallel. If we consider a massive sphere, $p^2/2M \to 0$, then the condition for the velocity below which dissipation is not possible is

$$v_c = \frac{\varepsilon}{p}. \tag{10.5}$$

In this approximation, dissipation via a phonon excitation first becomes possible when the speed of the sphere is larger than the phonon speed. Because, in superfluid helium, there are no single-particle excitations with $v = 0$, the lowest speed at which dissipation occurs is the propagation speed of the rotons. Experimentally, it is found that, below the speed $v_c \approx 30\,\text{m/s}$, the viscosity is equal to zero. This speed is slightly smaller than the rotons' speed, $v_R$.

## 10.3  Superfluid Helium Droplets

In Göttingen, Toennies and collaborators have developed a method with which well defined droplets of liquid helium can be produced. Helium gas cooled to 30 K is adiabatically expanded through a jet of $5\,\mu\text{m}$ diameter into a vacuum. Thus, a well defined beam of droplets is produced and the size of the droplets is determined by a time-of-flight method. From the many applications of such drop beams, we will select one: how many $^4$He atoms must there be in a drop to develop superfluidity?

An elegant method to demonstrate superfluidity is to study the rotational spectra of molecules inside the helium. When the droplets become superfluid, the molecules rotate freely, as in a vacuum, and the rotational lines are narrow. The rotational spectra of the molecules in droplets with less than 35 atoms do not show narrow lines. For larger droplets, the lines become stepwise clearer and, in droplets with 60 atoms, they are clearly there. The rotation spectra measured in molecules in droplets of liquid $^3$He do not display narrow lines. At temperatures around 1 K, $^3$He is a normal Fermi liquid.

## 10.4  Superfluid $^3$He

When there is an attraction between fermions, at sufficiently low temperatures, bound or quasi-bound states with bosonic properties arise. We will loosely refer to such states as Cooper pairs. In practice, one observes various superfluid phases of $^3$He at temperatures $T \leq 2.8\,\text{mK}$. The spin–spin interaction between the $^3$He nuclei is responsible for the formation of Cooper pairs. This interaction is attractive when both magnetic moments are parallel, i.e., when the total spin is $S = 1$. Because the total wave function must be antisymmetric, both atoms are in a relative $L = 1$ state. This implies that the wave function of the Cooper pairs, and hence the order parameter, has a tensor character. This interaction is, though, very weak.

$$V_{ss} = \frac{\alpha g^2 (\hbar c)^3}{4(M_p c^2)^2} \left\langle \frac{1}{r^3} \right\rangle \boldsymbol{\sigma} \cdot \boldsymbol{\sigma}'. \tag{10.6}$$

If we substitute the values $g = -1.9$ (the magnetic moment of $^3$He is $\mu = g\mu_N$ and $\mu_N = 3 \cdot 10^{-8}$ eV/Tesla is the nuclear magneton) and of $\langle 1/r^3 \rangle$, using the average separation between the helium atoms, $\approx 0.2$ nm, into (10.6), we obtain $V_{ss} \approx 10^{-11}$ eV. This is four orders of magnitude smaller than the temperature of the phase transition, $T \approx 2.8$ mK. At this temperature, the spin–spin interaction is negligible compared with thermal fluctuations. Superfluid $^3$He is a collective state, in which the magnetic moments of the Cooper pairs are organised in the total volume. In the superfluid state, the Cooper pairs are in the ground state. The binding energy of the total sample is the product of the number of Cooper pairs in the ground state, $N_{CP}$, and the binding energy of an individual pair, $V_{ss}$ (cf. 10.6). The total energy of the sample is larger than that of the thermal fluctuations. The magnetic field produced through the oriented $^3$He nuclei in the superfluid state is around 3 mT. If we take into account that $\mu_N/\mu_B \approx 1/2000$, then we see that the degree of orientation of the nuclear magnetic moments in superfluid $^3$He is readily comparable with the degree of orientation of the electrons in ferromagnet material.

## Literature

F. Pobell, *Matter and Methods at Low Temperatures* (Springer, Berlin, 1996)
J.P. Toennies, A.F. Vilesov, K.B. Whaley, Superfluid Helium Droplets. Phys. Today (2001)

# Chapter 11
# Metals – Quasi-free Electrons

*Tous les genre sont bon, lors le genre ennuyeux.*

Voltaire

Metals are built out of atoms that possess one, two or three weakly bound electrons. In the condensed state, these electrons are delocalised and move as nearly-free particles among the atoms. The interior electrons and the nucleus together form firmly bound positive ions, which are ordered in a crystal lattice. The motion of the external electrons in a periodic potential can be described by a modulated plane wave; in ideal crystals, electrons do not scatter. Scattering only takes place off crystalline defects and thermal oscillations. Hence, the electrons in metals may, to a good approximation, be described as a Fermi gas in a potential well.

We want to consider three aspects of metals that can be explained through the Fermi gas model: the binding of the atoms in the crystal, electrical conductivity and thermal conductivity.

## 11.1 Metallic Bond

### 11.1.1 Metallic Hydrogen

We first want to demonstrate the nature of the metallic bond via the example of metallic hydrogen. Hydrogen exists in the metallic state only at very high pressure, e.g., inside the planet Jupiter. In the laboratory, up to now it has only been possible, at a pressure of more than 140 GPa, to take a drop of liquid hydrogen into the liquid metal state for about 0.1 ms. This was demonstrated by a drastic increase in the electrical conductivity. Metallic hydrogen in solid form, on the other hand, has not

© Springer-Verlag GmbH Germany 2017
B. Povh and M. Rosina, *Scattering and Structures*,
Graduate Texts in Physics, DOI 10.1007/978-3-662-54515-7_11

**Fig. 11.1** The cells in a
cubic lattice are replaced by
spheres with radius $r_\text{s}$

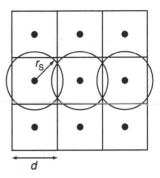

yet been created in the laboratory. It is suspected that pressures of around 500 GPa
will be needed for this.

We now want to sketch *on the back of an envelope* the metallic bond in hydrogen
under normal conditions.

Let us consider a closely packed cubic lattice of protons with a uniform distribution
of delocalised electrons. The crystal cell $d^3$ may be well replaced by a sphere of radius
$r_\text{s}$, such that the density, $N/\mathcal{V} = d^{-3} = (4\pi r_\text{s}^3/3)^{-1}$, stays the same (cf. Fig. 11.1). The
energy of the electrons in metallic hydrogen may be found using variational
methods and then compared with the energy in an isolated atom.

The average kinetic energy of the electron in a Fermi gas is from (9.8)

$$K = 2.21 \frac{\hbar^2}{2mr_\text{s}^2}, \tag{11.1}$$

where we have replaced the average atomic separation, $d$, by $r_\text{s}$.

The electrostatic energy of the proton at the centre of this sphere with constant
charge density $\rho = -3e/(4\pi r_\text{s}^3)$ is

$$V = \int_0^{r_\text{s}} -\frac{\alpha\hbar c}{r} \frac{3}{4\pi r_\text{s}^3} 4\pi r^2 \mathrm{d}r = -\frac{3}{2}\frac{\alpha\hbar c}{r_\text{s}}. \tag{11.2}$$

Neighbouring spheres do not contribute to this because they are neutral and spherically symmetric. Minimisation of the total energy, $E = K + V$, using (11.1) and
(11.2) leads to

$$r_\text{s} = 1.47\, a_0, \qquad E = -1.02\,\text{Ry}. \tag{11.3}$$

A more exact calculation (using modulated plane waves) yields $E = -1.05\,\text{Ry}$.
This energy should suffice to keep hydrogen atoms together through a metallic bond
(Fig. 11.2) because the binding energy of an electron in a free hydrogen atom is just
$E_\text{H} = -1\,\text{Ry}$. Nonetheless, hydrogen atoms at normal pressures do not form metals
because it is energetically preferable to form hydrogen molecules. The electron's

**Fig. 11.2** The charge distribution in a free hydrogen atom and in a hypothetical hydrogen metal at atmospheric pressure

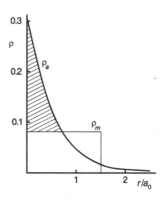

binding energy in a hydrogen molecule is, namely, $-1.17$ Ry per atom. This is why solid hydrogen at normal pressures is a crystal of molecules that are held together by the van der Waals force.

If one compares the charge distribution in our hypothetical hydrogen metal at standard pressures (Fig. 11.2) to that in the hydrogen molecule (Fig. 6.3), one sees that the electron density between the protons in molecules is significantly higher than in the metal.

At high pressures, the situation is different. When the separations between the hydrogen molecules become comparable with those inside the molecules, the electrons are no longer bound to particular electrons. Therefore, the metallic bond is stronger than the molecular one.

It is very different in metals for which the covalent bond between two atoms in a molecule is not as effective as in hydrogen or the majority of nonmetals. The condition for a metallic bond at normal pressure is thus that the binding energy of the delocalised electron gas is greater than the binding energy for individual molecules and not just for individual atoms.

### 11.1.2 Normal Metals

Our above estimate may be easily extended to the familiar metals. Consider, for example, sodium, which possesses a single conducting electron per atom. The main difference, compared with hydrogen, is the presence of the inner electron shells, which, because of the Pauli principle, keep the conducting electron away from the positive ion and act as a repulsive pseudo-potential. On the other hand, the potential inside the ion is larger than $e^2/(4\pi\varepsilon_0 r)$ and is only reduced by screening outside the ionic radius to $e^2/(4\pi\varepsilon_0 r)$. Both effects together may be simulated by a pseudo-potential, which is constant out to the ionic radius, $r_1$, and thereafter decreases as $-e^2/(4\pi\varepsilon_0 r)$ (Fig. 11.3). It is a good idea to fix the radius, $r_1$, so that the ionisation energy of the 3 s electron in a free Na atom corresponds to its experimental

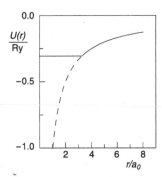

**Fig. 11.3** The pseudo-potential in sodium. The *dashed line* is the continuation of the Coulomb potential, $V = -e^2/(4\pi\varepsilon_0 r)$. Inside $r_s = 3.26a_0$, the actual potential is stronger because the inner shells are not completely screened; the Pauli principle, though, has a repulsive action and the two effects together are simulated by a constant pseudo-potential

value. For this, one finds, with numerical calculation, $E_{3s} = -0.378$ Ry by choosing $r_I = 3.26 a_0$.

Following our approach to hydrogen, one obtains

$$E = -\frac{3}{2}\frac{e^2}{4\pi\varepsilon_0 r_s} + \frac{1}{2}\frac{e^2 r_I^2}{4\pi\varepsilon_0 r_s^3} + 2.21\frac{\hbar^2}{2mr_s^2}. \tag{11.4}$$

The minimum energy is found at

$$r_s = 4.08 a_0, \quad E = -0.446 \, \text{Ry}. \tag{11.5}$$

The energy gain is thus calculated to be $\Delta E = E - E_{\text{atom}} = -0.068$ Ry $= -0.93$ eV. This coarse estimate is surprisingly close to the experimental values, $r_s = 4.00a_0$ and $\Delta E = -1.11$ eV.

**Fig. 11.4** The electron density, $\rho_a$, in a Na atom and $\rho_m$ in metallic Na. In metals, the electron density is shifted from the regions $r < r'$ and $r > r_s$ into the region $r' < r < r_s$

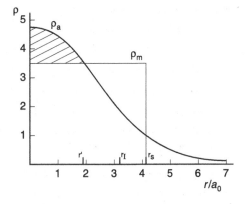

The essence of the metallic bond is the same as with the covalent bond. The periphery of the electron wave function at $r > r_s$ (Fig. 11.4) is attributed to the neighbouring atoms' cells. Because the electron distribution is thus brought a bit closer to the positive nucleus, the gain in the potential energy is larger than the loss from a generally smaller increase in kinetic energy (Fig. 11.4).

## 11.2 Electrical Conductivity

Electrical conductivity in metals may be well described under the assumption that the outermost electrons (the Fermi gas of electrons) move evenly under the influence of an electric field with a drift velocity $v_D$. In an electric field, electrons are accelerated, $dv/dt = (e/m)\mathbf{E}$. This acceleration is, though, only effective during the period $\tau$ between two electron–electron or electron–phonon collisions. The drift velocity of the electrons is thus

$$v_D = (e/m)\, E\, \tau .\tag{11.6}$$

The electrons are in a degenerate state, thus only those in the neighbourhood of the Fermi surface are scattered. These electrons move with the Fermi velocity, $v_F$. If we denote the mean free path by $\bar{l}$, the average time between two collisions is $\tau = \bar{l}/v_F$. This implies the current density

$$j = env_D = (ne^2\tau/m)\, E = \sigma E \tag{11.7}$$

and electrical conductivity

$$\sigma = \frac{ne^2\tau}{m} = \frac{ne^2\bar{l}}{mv_F}, \tag{11.8}$$

where $n$ is the number of conducting electrons per unit volume. In sodium and copper, one electron per atom participates in metallic bonding and electrical conductivity. For copper, from the measured conductivity, one can estimate $\tau \sim 7 \cdot 10^{-14}$ s and $\bar{l} \sim 30$ nm. The mean free path, $\bar{l}$, in copper is around 100 times larger than the separation between atoms.

## 11.3 Cooper Pairs

At low temperatures, many metals are superconducting and the electrical resistance vanishes. The mechanism of superconductivity is qualitatively well understood. Electrons at the Fermi surface, which at normal temperatures contribute to resistance through their scattering off crystal defects and thermal oscillations, bind at low temperatures in Cooper pairs. These behave like bosons and enter a Bose condensate with an energy gap.

The superconducting current is understood as a collective motion of Cooper pairs, which, because of the energy gap, are prevented from scattering. Here we want to concern ourselves with the question of how an effective, attractive potential is created in metals to bind the electrons in Cooper pairs.

The properties of the crystal lattice are highly dependent on the ratio of the electron mass, $m$, and the ion mass, $M$. For metals ($M \approx 50\,\mathrm{u}$), we have $M/m \approx 10^5$ and so $\sqrt{M/m} \approx 300$. Because ions and electrons in such a crystal with lattice separation $d$ are exposed to a similar force, $F \sim \alpha\hbar c/d^2$, their frequencies are inversely proportional to the square roots of their masses,

$$\frac{\omega_D}{\omega_e} \approx \sqrt{\frac{m}{M}}\,. \tag{11.9}$$

For the ionic frequency, we have taken the Debye frequency of the crystal, $\omega_D$; the electron frequency corresponds to the binding energy of the valence electron in the atom, $\hbar\omega_e = E_e \sim \alpha\hbar c/d$. The speed of sound and electron speed are related by the same ratio,

$$v_{\text{phonon}} \sim d\,\omega_D, \qquad v_e \sim d\omega_e, \qquad \frac{v_{\text{phonon}}}{v_e} \sim \sqrt{\frac{m}{M}}\,. \tag{11.10}$$

The electron speed ($10^6$ m/s) is indeed 300 times larger than the speed of sound in metal (3000 m/s).

When an electron flies past an ion, it transfers to it (see Fig. 11.5) momentum,

$$p = F\tau \approx \frac{\alpha\hbar c}{d^2}\frac{d}{v_e} \approx \frac{E_e}{v_e}, \tag{11.11}$$

where $\tau = d/v_e$ is the time that the electron spends in the vicinity of the ion. The ion then carries out a single oscillation with amplitude

$$\delta = \frac{p}{M\omega_D} \sim \frac{E_e/v_e}{\sqrt{Mm}\,\omega_e} \sim \frac{E_e}{mv_e^2}\sqrt{\frac{m}{M}}\,d \sim \sqrt{\frac{m}{M}}\,d, \tag{11.12}$$

before then again returning to its original state with a relaxation time $\omega_D^{-1}$. At temperature $T \sim 0\,\mathrm{K}$, the electrons cannot excite the lattice atoms because inelastic scattering can only take place off thermal fluctuations. Inside the relaxation time, the

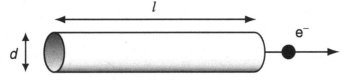

**Fig. 11.5**  The cylinder behind an electron in which the crystal lattice is distorted and the ions are pulled along in the direction of the axis

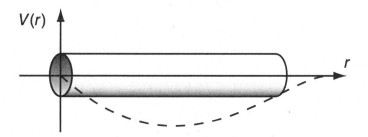

**Fig. 11.6** The perturbation of a crystal in a cylinder creates an attractive potential for other electrons (*dashed curve*)

electron proceeds through a distance $l \sim v_e(1/\omega_D) \sim \sqrt{M/m}\, d$. Thus, the perturbation of the crystal lattice is contained inside a cylinder with diameter $d$ and length $l$ (Fig. 11.6). If the ions have approached the axis by $\delta$, an attractive potential is created for further electrons. To exploit such a potential, the second electron must fly straight through the cylinder. The cylinder is not fixed in space because we are dealing with an s wave state. To obtain the zero angular momentum of the Cooper pair, its wave function must be a superposition of cylinder states in all directions.

Any net angular momentum, $\ell\hbar$, would keep the electrons $R \sim \ell\hbar/p \sim \ell d$ apart and so outside our cylinder of width $d$. Thus, Cooper pairs have zero spin and angular momentum. (Because an s wave is symmetric, the Pauli principle demands that the spin wave function is antisymmetric.)

The change in the potential due to the polarisation of the lattice is proportional to $\delta/d$, so that the attractive potential for the electrons in a relative s state takes the following form:

$$V(r) \begin{cases} \sim \dfrac{\delta}{d} \cdot \dfrac{\alpha\hbar c}{d} & \text{for } r < l \\[2mm] \approx 0 & \text{for } r > l, \end{cases} \tag{11.13}$$

where $l \approx \sqrt{M/m}\, d \approx 300d$. This is a relatively strong potential. It is not deep but has a large extension, so that one could expect that the binding energy roughly corresponds to the potential's depth, $E_e/300 \approx 3 \cdot 10^{-3}$ eV. One can easily convince oneself of this from the solution of the Schrödinger equation for the potential (11.13). Because this binding energy corresponds to a temperature of circa 30 K, one might expect many superconductors at reasonably high temperatures. In reality, the binding energy of Cooper pairs is always around $10^{-4}$ eV. Where then is the error in our reasoning?

Cooper pairs are formed from electrons just above the Fermi surface. The states below it are occupied and can thus not contribute anything to the wave function of Cooper pairs, which would correspond to a highly excited state in the potential (11.13). To illustrate the binding in Cooper pairs, let us compare the wave function of two free electrons (Fig. 11.7) with that of two electrons bound in a Cooper pair (Fig. 11.8). The wavelength of the electrons at the Fermi surface is much smaller

**Fig. 11.7**  The wave function
of two noninteracting
electrons. $\lambda_F$ is the
wavelength of an electron at
the Fermi surface

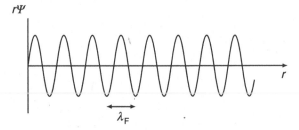

**Fig. 11.8**  The wave function
of a Cooper pair. The
coherence length is a few
hundred lattice separations
and $\lambda_F$ is the wavelength of
the electron at the Fermi
surface

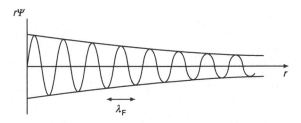

than the extension of Cooper pairs. Despite the weakness of the bond, it is enough
at low temperatures ($k_B T < E_{\text{binding}}$) to form a Bose condensate of Cooper pairs, so
many metals do become superconducting.

## 11.4   Diamagnetism in Superconductors

In high-temperature superconductors, it is very easy to demonstrate diamagnetism in
superconductors (the Meissner effect). A ceramic ring, with superconducting properties at the temperature of liquid nitrogen, is put into a container full of liquid nitrogen.
If one now puts the container into a magnetic field, the ring rises up and hovers.

Searching for analogies, we here use that between the diamagnetism of a noble gas
atom, such as neon, and the behaviour of a superconductor in a magnetic field – the
Meissner effect. A neon atom has total angular momentum $J = 0$; all the electrons
are coupled in pairs to angular momentum zero. If the neon atom enters a magnetic
field, then, from Lenz's rule, a magnetic moment opposed to the applied field is
induced. How, though, can a system with $J = 0$ have a magnetic moment? This is
explained as follows: as the magnetic field is switched on, an electric field is induced.
The electrons coupled in pairs move in opposed directions. One is accelerated by
the field and the other is braked. Thus, an electric current is produced and the entire
electron shell rotates around the axis of the applied magnetic field.

We will look at the superconductor for a one-dimensional geometry. In Fig. 11.9,
the surface of the superconductor is in the $yz$ plane, while the homogenous magnetic field outside the superconductor points in the $z$ direction. Let us ask how the
magnetic field and current behave inside the superconductor in the $x$ direction. In a

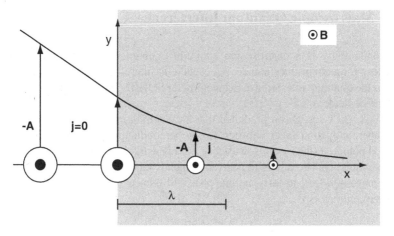

**Fig. 11.9** The surface of the superconductor is in the $yz$ plane, the homogenous magnetic field outside the superconductor points in the $z$ direction; it falls off inside with a range $\lambda$

magnetic field $\mathbf{B} = \nabla \times \mathbf{A}$, the kinetic energy is $(\mathbf{p} - e\mathbf{A})^2/2m$ and the velocity is then not $\mathbf{p}/2m$ but

$$\mathbf{v} = \frac{1}{m}(\mathbf{p} - e\mathbf{A}). \tag{11.14}$$

The current density $\mathbf{j}$ generated by our electron ensemble is thus

$$\mathbf{j} = e \sum_i \mathbf{v}_i = \frac{e}{m} \sum_i (\mathbf{p}_i - e\mathbf{A}). \tag{11.15}$$

Because $\sum \mathbf{p}_i = 0$, the proportionality of the current density and the vector potential follows:

$$\mathbf{j} = -\frac{ne^2}{m}\mathbf{A} = -\frac{1}{\lambda^2 \mu_0}\mathbf{A}. \tag{11.16}$$

In (11.16), $n$ is the number of electrons per unit volume and $\lambda = \sqrt{m/ne^2\mu_0}$ is the penetration depth of the magnetic field into the superconductor, as can be seen from the equation

$$\nabla \times \nabla \times \mathbf{A} = \mu_0 \mathbf{j} = -\frac{1}{\lambda^2}\mathbf{A}, \tag{11.17}$$

or, in our one-dimensional geometry,

$$\frac{\mathrm{d}^2 A_y}{\mathrm{d}x^2} = \frac{1}{\lambda^2}A_y. \tag{11.18}$$

The solution is an exponential decay, $A_y(x) = A_y(0)\exp(-x/\lambda)$.

## 11.5  Macroscopic Quantum Interference

Superconductivity is a macroscopic quantum phenomenon, which even permits macroscopic quantum interference. An especially interesting and significant application is the quantum interference magnetometer SQUID (superconducting quantum interference device).

In 1962, B.D. Josephson predicted that a tunnelling current could flow through a nonsuperconducting layer between two superconductors even when there is no electrical potential (the Josephson effect). For his work, he received the Nobel Prize unusually quickly. The tunnelling current depends on the critical current, $I_c$, which the Josephson bridge can carry and the phase difference, $\varphi = \varphi_1 - \varphi_2$, between the two superconductors is

$$I = I_c \sin \varphi. \tag{11.19}$$

The index $i = 1, 2$ refers to the two superconductors. The phases, $\varphi_i$, of the wave function of the Cooper pairs, $\psi_i$, have thus an important physical meaning, which we want to look at more carefully. In a magnetic field, the momentum of a Cooper pair (with charge $2e$ and effective mass $m^*$) is

$$\mathbf{p} = m^*\mathbf{v} + 2e\mathbf{A}. \tag{11.20}$$

One may write the wave function of the Cooper pair as $\psi(\mathbf{r}) = \exp(i\theta(\mathbf{r}))$ and express its momentum through the gradient of the phase, $\mathbf{p} = \langle \psi | - i\hbar\nabla | \psi \rangle = \hbar\nabla\theta$. The current density is thus

$$\mathbf{j} = 2en\mathbf{v} = (2en\hbar/m^*)\left(\nabla\theta - (2e/\hbar)\mathbf{A}\right). \tag{11.21}$$

Here, $n$ is the Cooper pair density. The current is then proportional to the gradient of the total phase, $\theta' = \theta - \int (2e/\hbar)\mathbf{A}d\mathbf{s}$. The phase gradient is thus a measurable physical quantity.

To derive equation (11.19), we follow an argument due to Feynman. The time development of the wave functions is essentially given by the Schrödinger equation of the two-state system, where both superconductors differ by a potential difference $(2e)V$ and are coupled by a tunnel integral, $K$,

$$i\hbar\frac{\partial\psi_1}{\partial t} = -eV\psi_1 + K\psi_2$$
$$i\hbar\frac{\partial\psi_2}{\partial t} = +eV\psi_2 + K\psi_1. \tag{11.22}$$

Expressing the wave functions by $\psi_i = \sqrt{n_i}\exp(i\varphi_i)$. we get

$$\frac{\partial n_1}{\partial t} = +\frac{2K}{\hbar}\sqrt{n_1 n_2}\sin\varphi \qquad (11.23)$$

$$\frac{\partial n_2}{\partial t} = -\frac{2K}{\hbar}\sqrt{n_1 n_2}\sin\varphi$$

$$\frac{\partial\varphi}{\partial t} = \frac{2eV}{\hbar}.$$

The current $I$ is proportional to $\partial n_1/\partial t$ and the proportionality coefficient $(2K/\hbar)\sqrt{n_1 n_2}$ in (11.23) corresponds to the critical current, $I_c$ in (11.19). Without the voltage $V$ the phase $\varphi$ is constant and thus the current is also constant. In a potential $V \neq 0$, the phase grows proportionally to the time, $\varphi = (2e/\hbar)Vt$, and one obtains an alternating current with frequency $\hbar\omega = 2eV$.

If one splits the current into two branches (each with its own Josephson bridge) and then brings them back together, the two currents sum up and interfere (Fig. 11.10a). For zero magnetic field, both are practically in the same phase and add constructively. In a magnetic field, each conductor acquires an additional phase $(2e/\hbar)\int \mathbf{A}\mathrm{d}s$, and

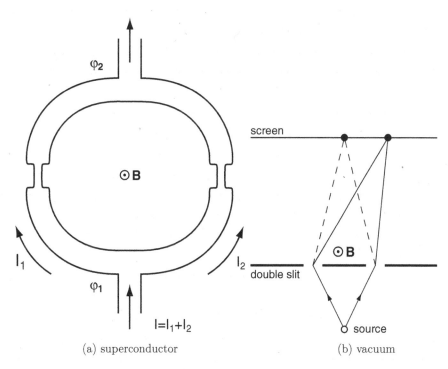

(a) superconductor (b) vacuum

**Fig. 11.10** (a) Principle of the SQUID magnetometer: the current is shared between two Josephson bridges and both contributions interfere. (b) Shift of the interference maxima in the Bohm–Aharonov effect; *dashed lines* without a magnetic field, *solid line* with a magnetic field

interference may be constructive or destructive. The vector potential **A** is not gauge invariant but one only needs the phase difference between the two branches, which is indeed gauge invariant. The integral over the loop yields

$$\varphi = \oint \frac{2eA}{\hbar} \, ds = \int \frac{2eB}{\hbar} \, dS = \frac{2e}{\hbar} \Phi = 2\pi \frac{\Phi}{\Phi_0}. \qquad (11.24)$$

Here, $\Phi$ is the magnetic flux in the loop and $\Phi_0 = 2\pi\hbar/2e = 2.07 \cdot 10^{-15}$ Vs is the magnetic flux quantum. One easily sees that the maximal current is produced when an integer multiple of the magnetic flux quantum is in the loop and the minimal one when a half-integer one is in it.

One can thus measure the magnetic field (or flux) by slowly increasing the field and counting the maxima and minima of the current: $\Phi = n\Phi_0$. Rather than the current, one measures in modern SQUID magnetometers the potential between both sides, which is produced keeping the current constant. The voltage arises when the current is hypercritical. The voltage is larger in the case of destructive than in the case of constructive interference; therefore, it changes between maxima and minima when the magnetic field is increased. We will not go into technical details here.

Such measurements are reminiscent of interferometric measurements of the width of a hair between two nonparallel glass plates; there, too, one only needs to count the interference lines.

The phenomenon in a SQUID is very analogous to the Bohm–Aharonov effect, where the electrons pass through two slits and interfere on a screen. If one puts a magnetic field between the two beams, then the interference lines are shifted (Fig. 11.10b) due to the additional phase difference, $\varphi = \oint (eA/\hbar) \, ds = \int (eB/\hbar) \, dS = (e/\hbar) \, \Phi$.

## 11.6   Thermal Conductivity

Thermal conductivity means energy transport and is produced in a gas by the movement of molecules or atoms, in nonmetals by phonons and in metals by electrons (and only to a very tiny extent by phonons). All these energy carriers are viewed, to a first approximation, as free particles. The interaction between these particles – and in the case of solids their interaction with crystal defects – are taken into account through their mean free paths, $\bar{l}$. It is illuminating to compare the thermal conductivity of gases, nonmetals and metals so as to understand the reason for the large differences in their thermal conductivities. The thermal conductivity of gases is so understood. Let the particle density in the gas be $n$ and the average speed $\bar{v}$. Then, through a surface $S$ from the left-hand (colder) side (see Fig. 11.11), $n\, \bar{v} \cos\theta\, S$ molecules per second pass from a distance $\bar{l}$ into the warmer region. Each of these molecules carries energy $c_v\, (T - (dT/dx)\, \bar{l})$. The same particle flux passes through $S$ from the right (warmer) side. The warmer particles carry, though, a larger energy, $c_v\, (T + (dT/dx)\, \bar{l})$ per molecule. With $\overline{\cos\theta} = 1/4$, the net heat flux is

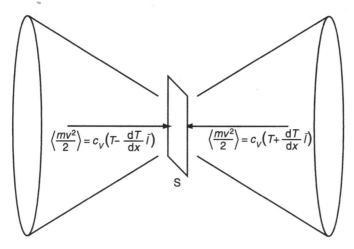

**Fig. 11.11** Energy transported through a surface $S$ by molecules, phonons or electrons

$$J_Q = \frac{1}{2} n \bar{v} c_v \bar{l} S \frac{dT}{dx}, \tag{11.25}$$

from which the thermal conductivity,

$$\lambda = \frac{1}{2} n \bar{v} c_v \bar{l}, \tag{11.26}$$

may be calculated. The precise numerical factor depends on the approximation used and the particular gas.

The same estimation scheme may be applied to phonon gases and electron gases. A comparison of the relevant factors is given in Table 11.1.

Due to the low density of air, its thermal conductivity is around 100 times smaller than that of nonmetals (Table 11.1). The high speed of electrons at the Fermi surface is again responsible for the roughly 100 times larger thermal conductivity of metals compared with nonmetals. The small $c_v$ value in copper refers, of course, only to the electrons.

Because the electrical and thermal conductivity in metals are caused by the electrons, it is sensible to consider them together. To this purpose, we require the explicit

**Table 11.1** Typical values of the factors in the calculation of the thermal conductivity for diverse materials in normal conditions. For phonons, only the product $nc_v$ is meaningful

| Medium | Carrier | $\bar{l}$ nm | $\bar{v}$ m/s | $n$ kmol/m³ | $c_v$ | $\lambda$ W/mK |
|---|---|---|---|---|---|---|
| Air | Molecules | 65 | 500 | 0.045 | 2.5 $R$ | 0.03 |
| Nonmetal | Phonons | 1 | 3000 | (45 | × 3 $R$) | 3 |
| Copper | Electrons | 30 | $10^6$ | 45 | 0.03 $R$ | 300 |

expression for the specific heat of a degenerate electron gas,

$$c_v = \frac{1}{2}\pi^2 k_B^2 T / \varepsilon_F \, , \tag{11.27}$$

which we will not derive here. From

$$\varepsilon_F \approx \frac{5}{3} m \bar{v}^2 / 2, \tag{11.28}$$

we find

$$c_v \approx \frac{1}{2} \frac{\pi^2 k_B^2 T}{\frac{5}{6} m \bar{v}^2} \, , \tag{11.29}$$

which fixes the Wiedemann–Franz ratio between the thermal and electrical conductivities,.

$$L_{W-F} = \frac{\lambda / T}{\sigma} \approx \frac{\frac{1}{2} n \bar{v} \bar{l} \frac{3}{5} \pi^2 k_B^2 / (m \bar{v}^2)}{n e^2 \bar{l} / (m \bar{v})} = \frac{3\pi^2}{10} \frac{k_B^2}{e^2} \tag{11.30}$$

An exact calculation gives almost the same result,

$$L_{W-F} = \frac{\pi^2}{3} \frac{k_B^2}{e^2} = 2.45 \cdot 10^{-8} \, \text{W}\Omega/\text{K}^2 \, , \tag{11.31}$$

which agrees well with the measured values. For copper, for example, this ratio at $273\,\text{K}$ is $L = 2.23 \cdot 10^{-8}\,\text{W}\Omega/\text{K}^2$.

# Literature

J. Clarke, SQUIDs, in *Scientifc American*, vol. 36, August 1994

R.P. Feynman, R.B. Leighton, M. Sands, *The Feynman Lectures on Physics*, vol. III (Addison-Wesley, Reading, 1965)

J.P. Ketterson, Song, Superconductivity (Cambridge University Press, Cambridge, 1999)

C. Kittel, *Introduction to Solid State Physics* (Wiley, New York, 1995)

W.J. Nellis, Making Metallic Hydrogen, in *Scientifc American* 60 (2000)

F. Pobell, *Matter and Methods at Low Temperatures* (Springer, Berlin, 1996)

M. Tinkham, *Introduction to Superconductivity* (McGraw-Hill, New York, 1996)

T. van Duzer, C.W. Turner, *Principles of Superconductive Devices and Circuits* (Elsevier, New York, 1981)

V.F. Weisskopf, Search for simplicity: the metallic bond. Am. J. Phys. **53**(10), 940–942 (1985)

V.F. Weisskopf, *The Formation of Cooper Pairs and the Nature of Superconducting Currents* (CERN, Geneva, 1979)

J.M. Ziman, *Principles of the Theory of Solids* (Cambridge University Press, Cambridge, 1979)

# Chapter 12
# Hadrons – Atoms of Strong Interaction

*Getretner Quark wird breit, nicht stark.*

Goethe

Hadrons are basic systems of the strong interaction, which may be pictured as the atoms of the strong interaction. Our principal interest concerns the structure of the nucleons, the building blocks of nuclei. The spectroscopic properties of the nucleons are interpreted as due to their being constructed from constituent quarks.

We will try to show a plausible relation between a constituent quark and the bare quark of quasi-elastic (deep-inelastic) scattering using the simplest possible model of chiral symmetry breaking (the Nambu–Jona-Lasinio model). We will dedicate more room than usual to this because it contains the basic properties of spontaneous symmetry breaking in particle physics, e.g. in the Higgs model.

## 12.1  Quarkonia

The effective forces between nonrelativistic constituent quarks can be extracted from the spectroscopy of quarkonia. Although for a given potential, the Schrödinger equation uniquely predicts the energy spectrum, the solution of the inverse problem is ambiguous. One finds, however, simple effective potentials that reproduce most energy levels rather well. In Fig. 12.1, we compare the excitations of positronium and charmonium. This is a reasonable comparison to make because both systems are made from a particle and its antiparticle, and both systems can, to a good approximation, be described nonrelativistically. Electrons are light, but the binding in positronium is weak, so their relative velocities are small. The heavy quarks have a sufficiently large mass such that they move slowly, despite the strong interaction. Because the

© Springer-Verlag GmbH Germany 2017
B. Povh and M. Rosina, *Scattering and Structures*,
Graduate Texts in Physics, DOI 10.1007/978-3-662-54515-7_12

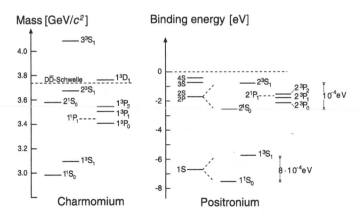

**Fig. 12.1** Charmonium and positronium states

interaction between quarks is transmitted by gluon exchange, which, like photons, are massless vector bosons, we expect that the potential between quarks – at least in the domain where one gluon exchange dominates – to behave as $1/r$.

The first thing we see is that the higher excited states of quarkonia do not lie closer together, as one would have expected for a pure Coulombic potential. We conclude that any description of quarkonia requires, as well as the "strong Coulombic potential", an additional confinement potential (see Sect. 3.3).

The second thing that can be seen, although we do not show it explicitly, is that the excitation energies in charmonium and bottomium are almost identical. The potential between the quark and antiquark is thus such that the excitation energies depend little or hardly at all on the quark mass. The mass of the quark is only visible in quarkonia in spin–spin splitting

Both of these properties may be well described through a combination of the strong Coulombic potential and the linear confinement potential,

$$E = \frac{\hat{p}^2}{2(m_q/2)} - \frac{4}{3}\frac{\alpha_s \hbar c}{r} + kr + U_0. \tag{12.1}$$

In (12.1), we have employed the reduced quark mass, $m_q/2$. The constant $U_0$ takes care of the zero point of the potential. The constituent masses are not uniquely determined and their choice affects the constant $U_0$.

Using the linear potential alone, one can analytically solve the eigenenergies and eigenfunctions (Airy functions); however, for the combined potentials, this can only be done numerically. We want to show that one can graphically – *on the back of an envelope* – obtain the experimental results through an interpolation of the Coulomb and oscillator potentials ($V = kr^2/2$). We have chosen the oscillator potential because its levels are equidistant and we will not need to do much work. The dimensionless energy scale unit we use is the difference between the 2S and 1S states and the zero point is chosen to be the ground state energy (1S). The results of the

**Fig. 12.2** The excited states of charmonium and bottomium compared with the spectra (*empty circles*) calculated for the three potentials, $V \propto 1/r, r, r^2$

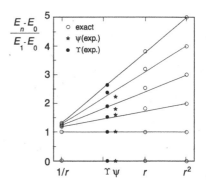

graphical solution are given in Fig. 12.2. The circles labeled "exact" are the results of the Schrödinger equation for the Coulomb, linear and quadratic potentials. One sees that the energy levels of the linear potential fit very well to the linear interpolation lines between the Coulomb and oscillator potentials. The levels of charmonium and bottomium lie nicely between the Coulomb and linear potential, which is compatible with a potential cocktail, as in (12.1).

One gets a good fit to the charmonium and bottomium spectra using the parameters $m_c = 1.37$ GeV $c_0^{-2}$, $m_b = 4.97$ GeV $c_0^{-2} 4.79$, $\alpha_s = 0.38$, $k = 0.860$ GeV fm$^{-1}$.

The strong coupling constant $\alpha_s = 0.38$ is clearly too big. From QCD, according to Eideman et al., at most a value of 0.2 could be expected in this energy region. This means that the fitted coupling constant, $\alpha_s$, represents an effective quark–quark interaction, which contains not just single gluon exchange but also other gluon field corrections. The fitted string constant $k = 0.860$ GeV fm$^{-1}$ is, on the other hand, too small, compared with the value $(1.0 - 1.2)$ GeV/fm, which one expects from lattice QCD or various phenomenological considerations. Both constants should be understood as effective values.

## 12.2 Hadrons from Light Quarks

The spectroscopic properties of baryons and mesons are very well described by a model that assumes that the u and d quarks have a mass of about 0.3 GeV $c_0^{-2}$, while for the s quarks, the value is around 0.5 GeV $c_0^{-2}$.

### 12.2.1 Nonrelativistic Quark Model

The masses of the light mesons are simply formed from the sum of the quark masses and spin–spin splitting,

$$M_{q\bar{q}} = m_i + m_{\bar{j}} + \Delta M_\sigma, \tag{12.2}$$

where

$$\Delta M_\sigma = \frac{8\pi\alpha_s \hbar^3}{9cm_i m_{\bar{j}}} |\psi(0)|^2 \langle \sigma_i \cdot \sigma_{\bar{j}} \rangle \,. \tag{12.3}$$

A good agreement with the measured meson masses is obtained using $m_{u,d} \approx 310 \,\mathrm{MeV}/c^2$ and $m_s \approx 483 \,\mathrm{MeV}/c^2$.

For the baryon masses, a similar relation holds,

$$M_B = \sum_i m_i + \Delta M_\sigma \,, \tag{12.4}$$

where the spin splitting depends on the relative orientation of the three quarks,

$$\Delta M_\sigma = \sum_{i<j} \frac{4\pi\alpha_s \hbar^3}{9cm_i m_j} |\psi(0)|^2 \langle \sigma_i \cdot \sigma_j \rangle \,. \tag{12.5}$$

The effective quark masses are found by fitting the calculated energy differences within a doublet with various spins to the measured baryon states. The best agreement is found with the quark masses $m_{u,d} \approx 363 \,\mathrm{MeV}/c^2$ and $m_s \approx 538 \,\mathrm{MeV}/c^2$. These masses are slightly different from those obtained from mesons. This is not surprising because the quarks in mesons and baryons inhabit different environments. In a meson, the quark couples to an antiquark, in a baryon to two quarks, which are coupled to the corresponding anticolour. The essential difference in the effective interaction is a factor of 2 (4/9 for barons, see (12.5), compared with 8/9 for mesons, see (12.3)), which is a consequence of the colour couplings. All other differences are taken into account through the different quark masses.

To keep dynamics out of the model, we have assumed that the potential and kinetic energies perfectly cancel each other. This is possible in a system in which the interaction between quarks may be described as the sum of a Coulomb and a linear potential. While for a Coulomb potential, $\langle E_{pot} \rangle = -2\langle E_{kin} \rangle$; in a linear potential, $\langle E_{pot} \rangle = 2\langle E_{kin} \rangle$. If the localisation probability of the quarks in both potentials are roughly the same, $\langle E_{pot} \rangle \approx -\langle E_{kin} \rangle$, the energy terms in the mass formula cancel each other. In fact, the localisation probability is larger for the linear potential and the negative constant $U_0$ in the potential (12.1) also helps produce the cancellation. The hadronic masses are, to a good approximation, reproduced by the sum of the quark masses and the spin–spin interactions – (12.2) and (12.4).

The constituent quark masses are certainly more than a simple numerical flourish of the model. One can convince oneself of this by comparing the predictions of the model for the baryon magnetic moments with the experimental results. The agreement is very good, if one assumes that the constituent quarks have the magnetic moments of a Dirac particle,

$$\mu_q = \frac{z_q e \hbar}{2m_q} \,. \tag{12.6}$$

**Table 12.1** Measured and calculated baryon magnetic moments in units of the magnetic moment of the nucleon, $\mu_N$. The experimentally determined magnetic moments of p, n and $\Lambda^0$ are used to calculate the magnetic moments of the remaining baryons. The $\Sigma^0$ hyperon is very short lived ($7.4 \cdot 10^{-20}$ s) and decays through the electromagnetic interaction via $\Sigma^0 \rightarrow \Lambda^0 + \gamma$. For this particle, instead of the expectation value of $\mu$, the transition matrix element $\langle \Lambda^0 | \mu | \Sigma^0 \rangle$ is quoted

| Baryon | $\mu/\mu_N$ (Experiment) | Quark Model | $\mu/\mu_N$ |
|---|---|---|---|
| p | $+2.792847386 \pm 0.63 \cdot 10^{-7}$ | $(4\mu_u - \mu_d)/3$ | – |
| n | $-1.91304275 \pm 0.45 \cdot 10^{-6}$ | $(4\mu_d - \mu_u)/3$ | – |
| $\Lambda^0$ | $-0.613 \pm 0.004$ | $\mu_s$ | – |
| $\Sigma^+$ | $+2.458 \pm 0.010$ | $(4\mu_u - \mu_s)/3$ | $+2.67$ |
| $\Sigma^0$ | | $(2\mu_u + 2\mu_d - \mu_s)/3$ | $+0.79$ |
| $\Sigma^0 \rightarrow \Lambda^0$ | $-1.61 \pm +0.08$ | $(\mu_d - \mu_u)/\sqrt{3}$ | $-1.63$ |
| $\Sigma^-$ | $-1.160 \pm 0.025$ | $(4\mu_d - \mu_s)/3$ | $-1.09$ |
| $\Xi^0$ | $-1.250 \pm 0.014$ | $(4\mu_s - \mu_u)/3$ | $-1.43$ |
| $\Xi^-$ | $-0.6507 \pm 0.0025$ | $(4\mu_s - \mu_d)/3$ | $-0.49$ |
| $\Omega^-$ | $-2.02 \pm 0.05$ | $3\mu_s$ | $-1.84$ |

The comparison between the experimental values and the predictions of the model is pretty good (see Table 12.1). The nonrelativistic quark model also correctly reproduces the order of magnitude of the excitations. The first excited state, with $\ell = 1$, lies at $\approx 0.6$ GeV.

## 12.3   Chiral Symmetry Breaking

In quasi-elastic scattering – known as deep-inelastic scattering in high-energy jargon – quark masses are estimated to be $m_q < 10$ MeV, which seems to contradict the nonrelativistic quark model with its much larger quark masses. A direct comparison is, however, inadmissible. In quasi-elastic scattering, we assume that we see the elementary, bare quarks. In low-energy experiments with poor resolutions, however, we see quarks surrounded by a cloud of gluons and quark–antiquark pairs. Our intuitive picture is as follows: in the case of a very weak interaction, the Dirac sea around the particle is undisturbed and its mass unchanged. If the strength of the interaction exceeds a critical value, the Dirac sea is very greatly disturbed and the particle dresses itself with many particle–antiparticle pairs. We call such a dressed particle, a quasi-particle and, in the strong interaction case, a constituent quark.

One tries to describe the relation between elementary and constituent quarks through a model of spontaneous chiral symmetry breaking. This symmetry breaking also represents, in today's cosmological scenarios, a link in the chain of phase transitions during the cooling of the universe. There is neither a discussion of this theme at an elementary level nor is it a topic in text books. We will therefore dedicate sufficient space to this theme, which appears to us to be conceptually very important, and will try to represent this somewhat complex phenomenon as simply as possible. A detailed treatment can be found in the review article by Klevansky.

If we were to bring the nucleons to a sufficiently high temperature and pressure, the constituent quarks would dissolve into their components – bare quarks. This reminds us of a phase transition. Phase transitions always involve spontaneous symmetry breaking during the transition to lower temperatures. In the quark case, it is the so-called chiral symmetry that is broken.

It is unfortunate that the procedure described above, which is so physically transparent, must be linked to an abstract symmetry to formulate it. The expression chirality or handedness is most familiar from optics. It is used to denote the properties of molecules that rotate the polarisation of light either to the left or to the right. In particle physics, chirality denotes a symmetry of the Dirac equation, which we wish to briefly explain. The reader can, though, without a great loss in understanding, jump directly from here to the constituent quarks (Sect. 12.3.1).

First, we want to explain the difference between chirality and helicity. Helicity is described by the operator $h = \boldsymbol{\sigma} \cdot \boldsymbol{p}/|\boldsymbol{p}|$, while chirality is described by the $\gamma_5$ (Dirac matrix) operator. Because, in relativistic quantum mechanics, fermions are described by four-component Dirac spinors, we need both quantum numbers to characterise the internal degrees of freedom.

Chirality is very clear for massless fermions, where it, as well as helicity, is a good quantum number. The Hamiltonian operator for a free fermion,

$$H = \gamma_0 \boldsymbol{\gamma} \cdot \boldsymbol{p}c \equiv \gamma_5 \, h \, |\boldsymbol{p}|c, \tag{12.7}$$

commutes with both the chirality operator, $\gamma_5$, and the helicity operator, $h$.

Because the electromagnetic as well as the weak and the strong interactions commute with $\gamma_5$, chirality is conserved in all processes. At high energies, for example, where the masses may be neglected, in $\beta$ decay, a left-handed fermion and a right-handed antifermion, $(e_R^+ + \nu_L$ or $e_L^- + \bar{\nu}_R)$, appear. The sum of the two chiralities before and after the decay is equal to zero.

If, though, a fermion has a finite mass, the mass operator $\gamma_0 mc^2$ does not commute with $\gamma_5$, and so chirality is no longer a good quantum number. One says that the mass breaks chiral symmetry. This property of chirality is used as a criterion to decide whether one is dealing with a massless or massive particle.

The wave functions of massless fermions provide, however, a basis for describing massive particles: one can decompose the wave function of a massive particle into two components, the right-handed and left-handed components, which correspond to right-rotating and left-rotating massless fermions. The right-rotating fermions and antifermions are right handed with probability $(1 + v/c)/2$ and left handed with probability $(1 - v/c)/2$ and vice versa for the left-rotating types.

Below, we want to show what happens with massless quarks whose coupling to virtual quark–antiquark pairs (the physical vacuum) preserves chirality but breaks the chiral symmetry of the vacuum as the coupling constant increases.

**Fig. 12.3** Multigluon exchange replaced by a contact interaction

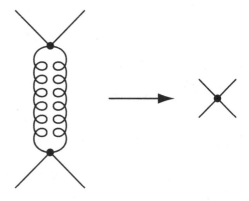

### 12.3.1 Constituent Quark

We want to simulate the phase transition, in which the massless quarks acquire a mass through spontaneous symmetry breaking, in a schematic model. The Nambu–Jona-Lasinio Model (NJL model) is suited to this, as it contains the essential low-energy property of QCD, chiral symmetry.

In this model the gluons are replaced by a contact interaction (Fig. 12.3), of which approximation is sufficiently good for low-energy hadronic physics. Here gluons never appear explicitly. In our simplified version of the NJL model, which we have pared down so that it can fit *on the back of an envelope*, we consider the strong interaction with just one quark flavour. We will merely qualitatively consider an extension of the model to the usual two quark flavours, up (u) und down (d).

The Hamiltonian in the simplified NJL model is given by

$$H = \int d^3r \left( -\bar{\psi} i\hbar c\gamma \cdot \nabla\psi + m_0 c^2 \bar{\psi}\psi \right)$$
$$- G \int d^3r \left[ (\bar{\psi}\psi)^2 + (\bar{\psi} i\gamma_5 \psi)^2 \right] . \tag{12.8}$$

The first line is the Hamiltonian for a free quark. The field operator, $\psi$, contains colour ($N_c = 3$), spin ($N_s = 2$) and also the four Dirac components. The second line simulates the chirality-preserving QCD interaction. The expression $(\bar{\psi} i\gamma_5 \psi)^2$ extends the contact interaction $(\bar{\psi}\psi)^2$ to form a chiral scalar.

The problem lies in finding the Hartree solution for a quark coupled to the vacuum fluctuations. The most general solution that we can expect is a plane wave that obeys $E = \sqrt{p^2c^2 + M^2c^4}$. The solution is found through the following trick: the effective quark propagator can be represented as a geometric series of graphs, as is symbolically shown in Fig. 12.4.

The trick needed for the solution lies in expanding the dressed propagator in a series of bare propagators, where we, though, use the dressed propagators in the loops. We seek a self-consistent solution of the equation

$$\mathcal{P}^{-1} = P^{-1} - A, \tag{12.9}$$

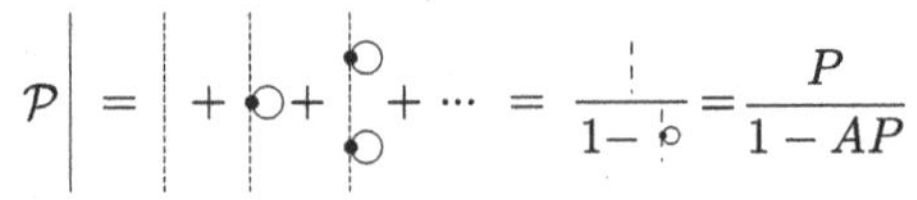

$$\mathcal{P}\Big| = \Big| + \big|\!\bigcirc\!+ \Big|\!\overset{\bigcirc}{\underset{\bigcirc}{\big|}}\!+ \cdots = \frac{\big|}{1- \bigcirc} = \frac{P}{1 - AP}$$

**Fig. 12.4** Constituent quark propagator in Hartree–Fock approximation. The *solid lines* represent the dressed propagators and the *dashes* the bare propagators. The bare propagator is $P$, the dressed one $\mathcal{P}$, and $A$ is the loop integral

which is symbolically represented in Fig. 12.4. If we take for the bare propagator

$$P = (\gamma^\mu p_\mu c + i\delta)^{-1} \tag{12.10}$$

and

$$\mathcal{P} = (\gamma^\mu p_\mu c - Mc^2 + i\delta)^{-1} \tag{12.11}$$

for the dressed one, we obtain

$$Mc^2 = A. \tag{12.12}$$

In this derivation, we have neglected the bare mass of the quarks, $m_0$. The value of the loop $A$ in Fig. 12.4 is the quark's self-energy! To emphasise that the bare quark has picked up a mass, we denote, in connection with chiral symmetry breaking, the constituent mass by $m_q = M$. The self-energy is given by summing over the internal degrees of freedom in the loop and integrating over the momentum $p$,

$$A = 2G \int_p \mathrm{Tr}\mathcal{P} = 2G \int_p \mathrm{tr}_C \mathrm{tr}_{\mathrm{Dirac}} \frac{1}{\gamma^\mu p_\mu c - Mc^2 + i\delta}. \tag{12.13}$$

Here, $2G$ is the value of the vertex in the Hamiltonian operator (12.8).

The evaluation of this expression is somewhat technical and represents a typical exercise for those that busy themselves with Feynman rules. The advantage of the Feynman rules is that one can picture the physical process and qualitatively understand the final result (12.17) even when one does not engage with the technical details. Through rationalising the fraction, the $\gamma$ matrices enter the numerator and the denominator develops terms quadratic in the momentum, $p_\mu$.

$$A = 2G \int_p \mathrm{tr}_C \mathrm{tr}_{\mathrm{Dirac}} \frac{\gamma^\mu p_\mu c + Mc^2}{p^\mu p_\mu c^2 - (Mc^2)^2 + i\delta}. \tag{12.14}$$

We use the traces $\mathrm{tr}_{\mathrm{Dirac}}\gamma^\mu = 0$ and $\mathrm{tr}_{\mathrm{Dirac}}1 = 4$. The sum over the internal degrees of freedom ($\mathrm{Tr} \equiv \mathrm{tr}_C \mathrm{tr}_{\mathrm{Dirac}}$) in the loop yields a constant factor, ($N_{Cs} = N_C \times N_s = 6$), and we only need to explicitly perform the integral over the four-momentum $cd^4 p$.

The integral over $p_0$ can be elegantly performed using Cauchy's theorem. One integrates, for example, over a contour that includes the lower pole in the complex plane ($p_0 = \sqrt{\boldsymbol{p}^2 + M^2 c^2} - i\delta$) and takes the residue

$$
\int_{-\infty}^{+\infty} \frac{-\mathrm{d}p_0}{2\pi i} \frac{2N_{\mathrm{Cs}} M c^2}{(p_0 - \sqrt{\boldsymbol{p}^2 + M^2 c^2} + i\delta)(p_0 + \sqrt{\boldsymbol{p}^2 + M^2 c^2} - i\delta)}
$$
$$
= -\frac{2N_{\mathrm{Cs}} M c^2}{2\sqrt{\boldsymbol{p}^2 + M^2 c^2}} \, . \tag{12.15}
$$

We are left with the three-momentum integration, which can be performed analytically; however, we will instead give a simple and instructive estimate. Because we want to describe low-energy hadronic excitations, we may stop the integration at a cut-off $\Lambda \approx 1\,\mathrm{GeV}$,

$$
A = 2G N_{\mathrm{Cs}} \int_0^\Lambda \frac{4\pi p^2 \mathrm{d}p}{(2\pi\hbar)^3} \frac{Mc}{\sqrt{p^2 + M^2 c^2}} \approx 2G N' \frac{Mc}{\sqrt{\bar{p}^2 + M^2 c^2}} \, . \tag{12.16}
$$

Here, $N' = N_{\mathrm{Cs}} \int_0^\Lambda 4\pi p^2 \mathrm{d}p/(2\pi\hbar)^3 = N_{\mathrm{Cs}} \Lambda^3/6\pi^2\hbar^3$ is the density of the momentum states multiplied by $N_{\mathrm{Cs}}$ and $\bar{p}$ is a suitable average value, around two thirds of the cut off, $\Lambda$.

The equation for the constituent mass (the "*gap equation*") thus has the form

$$
M = A/c^2 = 2G N' \frac{M}{c\sqrt{\bar{p}^2 + M^2 c^2}} \, . \tag{12.17}
$$

This always has a solution, $M = 0$. If, though, $(2G N')^2 > \bar{p}^2/c^2$, then there exists a further solution,

$$
(Mc^2)^2 = (2G N')^2 - (\bar{p}c)^2 \, , \tag{12.18}
$$

and indeed, this second solution with a finite mass minimises the energy of the system. One here sees clearly the phase transition as a function of the size of the coupling, $G$. Below a critical value of $G$, one only has the trivial solution $M = 0$, but above it is $M > 0$ (Fig. 12.5).

Let us recall here (5.27), which describes the phase transition to ferromagnetism, and compare it with (12.17) for the chiral phase transition. In both equations, the order parameters "$M$" (magnetisation or constituent mass, respectively) are associated with positive feedback.

The nonrelativistic quark model describes the ground state and Low-energy excitations of hadrons rather well. The masses of the Baryons, $M_B \approx 3m_q$, and of the mesons, $M_{q\bar{q}} \approx 2m_q$, are reproduced well by (12.4) and (12.2). The exception is the mass of the pion. This is a factor of five smaller than the mass of two quarks. In the constituent quark model, the strong spin–spin interaction is made responsible for

**Fig. 12.5** The dependence
of the constituent mass on
the strength of the coupling.
The cross corresponds to the
realistic values
$\Lambda = 0.631\,\mathrm{GeV}\,c^{-1}$,
$M = 0.335\,\mathrm{GeV}\,c^{-2}$

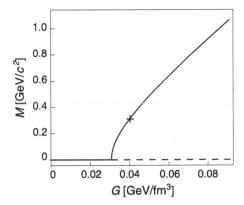

this mass decrease. This is indeed qualitatively correct. However, a nonrelativistic
calculation of such a large mass defect is nonsensical. We want to show in the framework of the NJL model that the small mass of the pion is a result of chiral symmetry
breaking.

### 12.3.2  The Pion

We can estimate the pion mass in our simplified model just as we did for the constituent quark mass. The pion has $J^\pi = 0^-$ and couples to quark–antiquark pairs
via a pseudoscalar interaction. In Fig. 12.6, we compare the pion propagator with its
expansion in quark–antiquark fluctuations. The left-hand side of Fig. 12.6 describes
resonant quark–quark scattering in terms of the pion,

$$(q\bar{q} \rightarrow \pi \rightarrow q\bar{q}) = -\mathrm{i}(\hbar c)^3 \frac{(\mathrm{i}\gamma_5 g_{\pi qq})(g_{\pi qq}\mathrm{i}\gamma_5)}{E^2 - m_\pi^2 c^4} , \tag{12.19}$$

while the right-hand side represents the microscopic description of the pion. The
pseudoscalar part of the contact interaction has the value

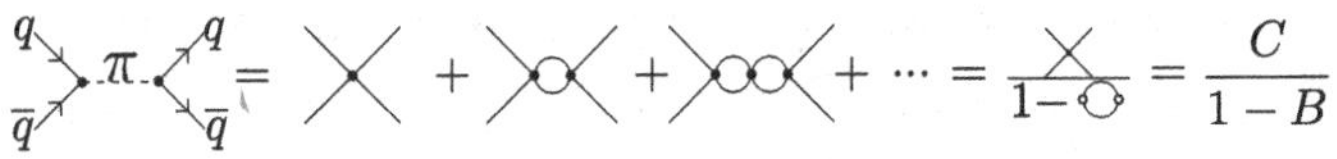

$$\underset{\bar{q}}{\overset{q}{\diagdown}}\pi\underset{\bar{q}}{\overset{q}{\diagup}} = \times \; + \; \rangle\!\!\times\!\!\langle \; + \; \rangle\!\!\times\!\!\times\!\!\langle \; + \cdots = \frac{\times}{1 - \bigcirc} = \frac{C}{1 - B}$$

**Fig. 12.6** The *left side* of the equation describes resonant quark–quark scattering via pion exchange,
while the *right* has the corresponding microscopic picture of the scattering using quark–antiquark
pairs coupled with the quantum numbers of the pion. The *dots* (*solid circles*) correspond to the
vertex $2G$, while the *empty circles* signify the factor $\sqrt{2G}\,\gamma_5$ because this vertex is associated to
two neighbouring loops in the geometric series

$$C = \; \times \; = -\mathrm{i} \cdot \mathrm{i}\gamma_5(-2G)\mathrm{i}\gamma_5 \,. \tag{12.20}$$

Equating the left- and right-hand sides of Fig. 12.6 yields the pion mass and the effective coupling constant, $g_{\pi qq}$,

$$(\hbar c)^3 \frac{g_{\pi qq}^2}{E^2 - (m_\pi c^2)^2} = \frac{-2G}{1 - B} \,. \tag{12.21}$$

One can read off the pion mass from the position of the pole ($B(E \to m_\pi) = 1$), and the numerator yields the coupling constant. The derivation of $B$ is analogous to that of the integral $A$. To understand the result, one just needs to realise that the loop $B$ contains two dressed quark propagators, where there was just one in $A$. The derivation is simple in the case where $E = 0$ and if one works in the rest frame of the pion ($p_\pi = 0$). Then, for the second quark propagator, $p_2^\mu = -p_1^\mu$ and one has, in the denominator, $(\gamma^\mu p_\mu c - Mc^2 + \mathrm{i}\delta)\gamma_5(-\gamma^\mu p_\mu c - Mc^2 + \mathrm{i}\delta)\gamma_5 = (\gamma^\mu p_\mu c - Mc^2 + \mathrm{i}\delta)^2$. After rationalizing the denominator, taking the trace over the $\gamma$ matrices and discarding the vanishing quadratic pole, it is found that the integrals $A$ and $B$ (for $E = 0$) only differ by a factor of $Mc^2$,

$$B(E^2 = 0) = 2G\mathrm{i} \int \frac{\mathrm{d}^4 p}{(2\pi)^4 \hbar^3 c} \mathrm{Tr} \frac{1}{p^\mu p_\mu - M^2 c^2 + \mathrm{i}\delta} = \frac{A}{Mc^2} = 1 \,, \tag{12.22}$$

which, up to the factor $Mc^2$, agrees perfectly with the expression in (12.16). For perfect chiral symmetry (quark mass $m_0 = 0$), one indeed finds the pole $[1 - B(E^2 = 0)]^{-1} = \infty$ at zero energy, which corresponds to a massless pion.

### 12.3.3 Generalisation to $m_0 \geq 0$ and Two Quark Flavours

The up and down quarks have, though, a small mass, around 2% of the constituent mass, and so chiral symmetry is slightly broken from the start. Therefore, also the constituent masses of the up and down quarks are slightly different. This difference is, though, tiny – it is comparable with that of the bare quarks.

In chiral symmetry breaking, one has to take two effects into account: explicit symmetry breaking due to the mass term $m_0 \neq 0$ in the Hamiltonian (12.8) and spontaneous symmetry breaking due to the particular interaction. The latter gives a much larger contribution ($A$) to the constituent mass ($Mc^2 = m_0 c^2 + A$) than the explicit term ($m_0 c^2$).

Explicit chiral symmetry breaking is much more important for the determination of the pion mass than it is for the quark masses. If there were only spontaneous symmetry breaking, the pion would be an exact Goldstone boson and its mass would vanish. Due to explicit chiral symmetry breaking from the finite mass of the bare quarks ($m_0 \neq 0$), the pion mass, while small, is nonzero.

Because pions are built from quarks with two flavours, we have not one but an isospin triplet of pions.

### 12.3.4  The Pion as a Collective State

One can develop a deeper insight into the special character of the pion and thus into the mechanism that produces Goldstone bosons, if one views it as a collective vibrational state of particle-hole states (quark antiquark pairs). The model we describe here is very similar to the model of giant resonances in nuclear physics (Chap. 14) and to the model of localised vibrational modes in crystals (Chap. 8). While in the shell model of nuclear physics one excites a nucleon from a filled to the valence shell, one can, in hadronic physics, promote a quark from the Dirac sea to the Fermi sea (Fig. 12.7). It should be noted that our particles are constituent quarks and the antiparticles are holes in the Dirac sea of constituent quarks. The collective state is then formed from a superposition of many particle–hole (quark antiquark) configurations, $\phi_i$,

$$|\Phi\rangle = \sum_{i=1}^{\tilde{N}} c_i |\phi_i\rangle . \qquad (12.23)$$

The index $i \equiv (p, c, f, s)$ here denotes the momentum, colour, flavour and spin components of the quarks as well as the opposite values for the antiquarks, so that the pion is coupled to zero momentum, colour and spin. The number of configurations is $\tilde{N} = N'\mathcal{V}$, where $N'$ is, as in (12.16) and (12.18), the number of quantum states per unit volume and the normalisation volume $\mathcal{V}$ ensures the correct dimensions.

**Fig. 12.7** Quark excitation from the Dirac sea into the Fermi sea

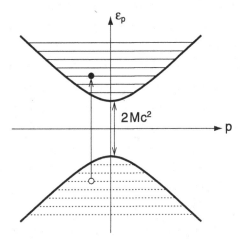

From this ansatz, we can solve the Schrödinger equation $H|\Phi\rangle = E|\Phi\rangle$. The coefficients $c_i$ fulfill the secular equation,

$$
\begin{pmatrix}
E_1 - 2\tilde{G} & -2\tilde{G} & -2\tilde{G} & \cdots \\
-2\tilde{G} & E_2 - 2\tilde{G} & -2\tilde{G} & \cdots \\
-2\tilde{G} & -2\tilde{G} & E_3 - 2\tilde{G} & \cdots \\
\vdots & \vdots & \vdots & \ddots
\end{pmatrix}
\cdot
\begin{pmatrix}
c_1 \\ c_2 \\ c_3 \\ \vdots
\end{pmatrix}
= E \cdot
\begin{pmatrix}
c_1 \\ c_2 \\ c_3 \\ \vdots
\end{pmatrix}. \tag{12.24}
$$

The diagonal elements contain the unperturbed energy of the quark–antiquark pair, $E_i = 2\sqrt{(Mc^2)^2 + (p_i c)^2}$, and the interaction in all diagonal and off-diagonal elements is, from (12.20), equal to $-2\tilde{G} = -2G/\mathcal{V}$, where the normalisation volume again ensures the correct dimensionality.

To emphasise the analogies between different areas of physics, we will apply the same schematic formalism to the description of the collective states not only in this chapter but also in Chap. 9 (localised vibrational modes) and Chap. 14 (giant resonances in nuclei). To solve the secular Equations, we express the coefficients $c_i$ in the diagonal elements as a sum of all the other coefficients,

$$
c_i = \frac{-2\tilde{G}}{E - E_i} \sum_{j=1}^{\tilde{N}} c_j, \tag{12.25}
$$

where $\sum_j c_j$ is a constant. Summing both sides over all $\tilde{N}$ quark–antiquark states and taking $\sum_i c_i = \sum_j c_j$ into account, the solution of the secular equation yields the relation

$$
1 = \sum_{i=1}^{\tilde{N}} \frac{-2\tilde{G}}{E - E_i}. \tag{12.26}
$$

It is best to represent the solution of this equation graphically (Fig. 12.8). The right-hand side of (12.26) has poles at the $E = E_i$ values. The solutions $E_i'$ are found where the right-hand side is equal to unity. These are marked on the abscissa. The $(\tilde{N} - 1)$ eigenvalues are trapped between the unperturbed energies, $E_i$. The outlier, marked as $E_\pi$, is the collective state (pionic ground state). For an attractive interaction, the collective state lies below quark–antiquark states.

To obtain a quantitative estimate of the energy shift, we assume that all states are degenerate, i.e., the energies $\bar{E}_i$ are the same for all $i$ and use an average momentum $\bar{p}$. Then, from (12.18), we obtain

$$
\bar{E}_i = 2\sqrt{(Mc^2)^2 + (\bar{p}c)^2} = 4GN', \tag{12.27}
$$

and the energy of the collective state is

$$
E_\pi = 4GN' - \tilde{N} \cdot 2\tilde{G} = 4GN' - N' \cdot 2G = 2GN'. \tag{12.28}
$$

**Fig. 12.8** Graphical
depiction of the solution of
the secular equation for the
pion. The $E_i$ values are
unperturbed energies, the $E_i'$
values are the diagonalised
energies and $E$ is the energy
of the collective state (pionic
ground state, $E = m_\pi c^2$)

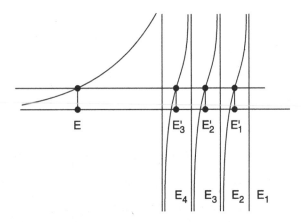

The masses of normal quarkonia lie around or above double the constituent quark mass, $M$. The pion is an exception. Due to collective effects, its mass drops, in our approximation from $4GN'$ to $2GN'$, so that the pion mass would be around 300 MeV.

This drop is not yet enough. One has to take correlations of the quarks in the ground state (vacuum) into account. Then the Fermi sea is partly populated by quarks and one can "de-excite" a quark from the Fermi into the Dirac sea. In the expansion of the collective state, the number of configurations is doubled since as well as $\phi_i$ also the corresponding de-excitations $\bar{\phi}_i$ occur.

$$|\Phi\rangle = \sum_{i=1}^{\tilde{N}} c_i |\phi_i\rangle + \bar{c}_i |\bar{\phi}_i\rangle . \tag{12.29}$$

In the so-called *random phase approximation* one indeed finds a secular equation that is twice as large as (12.24) and the pion rest energy drops to zero,

$$E_\pi = 4GN' - 2\tilde{N} \cdot 2\tilde{G} = 4GN' - 2N' \cdot 2G = 0. \tag{12.30}$$

(We skip certain technical details.) We have thus tried to indicate the equivalence of these two views of the pion – as a collective state and as a Goldstone boson.

For perfect chiral symmetry – in agreement with Goldstone's theorem – the continuous global symmetry is spontaneously broken and there exists a *soft mode* with eigenfrequency zero. If, though, the bare mass $m_0 \neq 0$, then chiral symmetry is explicitly broken and the pion acquires a finite, though small, rest mass ($E_\pi = m_\pi c^2 = 140$ MeV).

One can envisage the pion as a classical oscillation in a potential that describes the vacuum solution of the NJL model as a function of the order parameter, $M e^{i\phi}$ (Fig. 12.9). Here $M$ is the constituent quark mass and $\phi$ an arbitrary phase. The pion (the *soft mode*, $\hbar\omega \to 0$) corresponds to the oscillation along the bottom, i.e., along

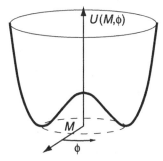

**Fig. 12.9** The representation of the pion and the $\sigma$ meson as natural oscillations in a potential with spontaneously broken chiral symmetry. This is known as a "Mexican hat" potential. The pion (the *soft mode*, $\hbar\omega \to 0$) corresponds to the oscillation along the phase angle $\phi$ and the $\sigma$ meson to the oscillation in the steep direction orthogonal to the ditch ($\hbar\omega \approx 2Mc^2$)

the phase angle, $\phi$, and the $\sigma$-meson corresponds to oscillations in the steep direction orthogonal to the bottom ($\hbar\omega \approx 2Mc^2$).

The question remains why only the pion is a Goldstone boson (and to a certain extent, for three light quarks, the kaon too). Because of chiral symmetry, the Hamiltonian in the NJL model contains two terms. The first is responsible for the constituent quark mass and does not generate a collective state. It represents the interaction between quark–antiquark pairs with quantum numbers $0^+$. Such a state with zero energy does not represent a new independent state and is identical to the vacuum. The second term represents the interaction of quark–antiquark pairs with quantum numbers $0^-$, and this generates the collective state: the pion.

## Literature

S. Eidemann et al., Phys. Rev. Lett. B **592** (2004)

S.P. Klevansky, The Nambu–Jona-Lasinio model of quantum chromodynamics. Rev. Mod. Phys. **64**, 649–708 (1992)

B. Povh et al., *Particles and Nuclei* (Springer, Berlin, 2015)

# Chapter 13
# The Nuclear Force – Pion Sharing

*So far as the laws of mathematics refer to reality, they are not
certain. And so far as they are certain, they do not refer to reality.*

Einstein

Continuing our attempt to bring out the principal contents of physics via analogies,
it is natural in the case of the nuclear force to use the analogy with the interatomic
force. Indeed, the nucleon–nucleon potential closely resembles that between two
atoms – if we decrease the length scale by around five orders of magnitude (from
$0.1\,\mathrm{nm} \rightarrow 1\,\mathrm{fm}$). When viewed in this way, the nuclear force is, though, weak, at
least when it is compared with the most important chemical bond, the covalent bond.
After all, while the chemical bond at low temperatures produces a solid state, nuclei
remain liquid even at a temperature $T = 0\,\mathrm{K}$.

The nucleon–nucleon interaction is best investigated by scattering. This interaction can be directly used to describe light nuclei where many body effects do not yet
dominate. This means that we can view the deuteron, tritium and the helium isotopes
as the molecules of the strong interaction.

In contrast with this, we are better off viewing nuclei with more than 16 nucleons
as droplets of a degenerate Fermi liquid. The interactions between nucleons in heavy
nuclei are primarily described by a common nuclear potential with a residual interaction, which is only qualitatively similar to the more fundamental nucleon–nucleon
interaction.

The nucleon–nucleon interaction has been investigated in detail with the help of
scattering at energies below the pion threshold. In this energy domain, in which only
elastic scattering is possible, the interaction may be well described by a local potential. The form of the potential is strongly dependent on the spin, isospin and orbital
angular momentum. In Fig. 13.1, we show (dashed lines) how large the repulsive and

© Springer-Verlag GmbH Germany 2017
B. Povh and M. Rosina, *Scattering and Structures*,
Graduate Texts in Physics, DOI 10.1007/978-3-662-54515-7_13

**Fig. 13.1** The
nucleon–nucleon potential.
Depending on the spin,
isospin and orbital angular
momentum, the potential can
be repulsive or attractive.
The bounds on the strength
of the potential are given as
*dashed lines*. The *solid line*
shows a potential that is
averaged over spin, isospin
and angular momentum

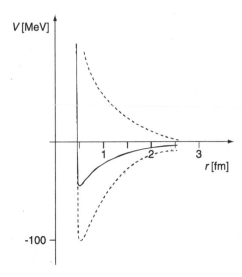

attractive potentials can be in different relative nucleon–nucleon states. The solid line
shows an averaged potential. Common to all the potentials is a repulsive interaction
at separations shorter than $r \approx 0.5$ fm.

## 13.1   Repulsion at Short Distances

As with atoms, short-distance nucleon repulsion is a consequence of the Pauli principle. In the lowest state with $\ell = 0$ one can, in principle, squeeze in 12 light quarks
(three colours, two flavours and two spins); however, an antisymmetric wave function for six quarks is orthogonal to the wave function for two overlapping nucleons.
At short distances between nucleons, some of the quarks are excited into higher
states and some of their spins are flipped. In both cases, energy is needed. A coarse
estimate yields that either exciting two quarks from the s to the p state or flipping
the quarks' spins ($2N \rightarrow 2\Delta$, i.e., the simultaneous excitation of two nucleons into
a two $\Delta$ excited states) each cost around 600 MeV. Both effects thus contribute to
short-distance repulsion. Overall, the compromise between all effects (minimisation
of the total repulsive energy) is around 300 MeV.

## 13.2   Attraction

In 1935 Yukawa formulated the theory of the interaction between the nucleons in
analogy to the electromagnetic interaction. Because of the short range of the nuclear
interaction the massless photon of the electromagnetic interaction was replaced by

an about 200 MeV massive pion. To explain the strength of the nuclear force the coupling constants of the pion and heavier exchange mesons to the nucleon are three orders of magnitude higher than the coupling constant of the photon to the electron. In the textbooks of the nuclear physics the nuclear interaction is illustrated using the classical field equation for mesons. This is analogous to the Poisson equation in electrostatics, however, with an additional term that takes into account the mass of the exchange particle (13.1).

$$\nabla^2 \Phi - (m_\Phi c/\hbar)^2 \Phi = -g\rho(\mathbf{r}) = -g[\delta(\mathbf{r}) + \delta(\mathbf{r} - \mathbf{R})]. \tag{13.1}$$

The solution of (13.1) is (with the range $b = \hbar/(m_\pi c) = 1.4\,\text{fm}$)

$$\Phi = -g\frac{e^{r/b}}{4\pi r} - g\frac{e^{|\mathbf{r}-\mathbf{R}|/b}}{4\pi |\mathbf{r} - \mathbf{R}|}. \tag{13.2}$$

The potential energy is calculated as in electrostatics,

$$V_{\text{pot}} = \frac{1}{2} \int g\rho(\mathbf{r})\Phi(\mathbf{r})\mathrm{d}^3 r = V_1 + V_2 + V(R). \tag{13.3}$$

Here, $V_1$ and $V_2$ are the $R$ independent contributions – the self-energies of the first and second nucleon – and $V(R)$ is the famous Yukawa potential,

$$V(R) = -g^2\frac{e^{-R/b}}{4\pi R}. \tag{13.4}$$

The simple form of this potential is only correct for scalar mesons such as, e.g., the $\sigma$ meson. The dominant contribution to attraction is indeed due to $\sigma$ meson exchange because this is independent of spin and isospin. For pions (pseudo-scalar charged mesons), the form of the potential is more complicated; it is spin and isospin dependent. All of these properties have been thoroughly confirmed experimentally.

## 13.3  Information from Light and Heavier Nuclei

We will treat the nuclear interaction as analogous to the inter-atomic interaction. The strong interaction is the interaction between the quarks and gluons. The nuclear interaction is analogous rather to the interaction between the atoms than to the elementary interaction between the electrons. The two interactions, the nuclear and the one between the atoms, display similar features, repulsion at small distances when they strongly overlap as the consequence of the Pauli principle and attraction when their surfaces touch. Consequently the potentials between nucleons and between atoms have similar form, strong repulsion by overlapping of the two and a short range attraction. In the deep inelastic scattering of electrons on protons it was found that not only quarks but also pions are the constituents of the proton. The proton is 25% of time a nucleon plus a pion and about the same fraction of time a proton plus a scalar boson. In the spirit of the Nambu–Jona-Lasino model of the chiral symmetry breaking we identify the pion with the Goldstone pion and the scalar boson with the $\sigma$ boson. The $\sigma$ boson has a large overlap with the resonance in the two pion scattering state.

It is remarkable that for nuclei above oxygen the volume and the total binding energy are proportional to the number of nucleons (the mass number $A$).This shows that the nuclear force is rather weak and that it acts only between nearest neighbours. This is related to the fact that the distance between neighbours is about 2 fm while the potential minimum is at about 1 fm. The binding energies above $A = 16$ have approximately the same value about 8 MeV (Fig. 13.2). In this region, the two pions coupled to the zero spin dominate the coupling. Analogous to the sharing of electrons between the atoms the pions spread in the nucleus.

The contribution of sharing one pseudo-scalar pion is more complicated. The pion of the donor nucleon as well as of the acceptor nucleon has the orbital angular momentum $l = 1$. The resulting effective force between the nucleons depends on their spins and isospins and is the source of the tensor force. In the heavy nuclei the contribution of one pion to the binding is believed to average out. In light nuclei, however, it may be the dominant binding force. This is demonstrated in several odd–odd nuclei, deuteron, $^6$Li and $^{14}$N. Deuteron is a bound proton and neutron, $^6$Li is a $^4$He core plus a bound proton and neutron, and $^{14}$N is a $^{16}$O core with a proton and neutron holes. All these three nuclei have a $J^\pi = 1^+$ ground state and a $J^\pi = 0^0$ first

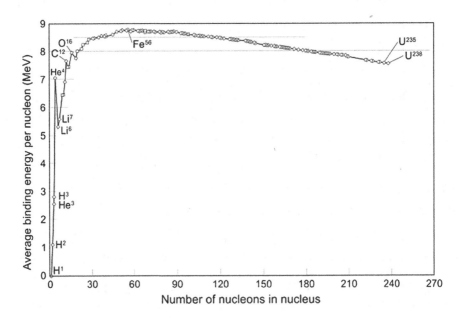

**Fig. 13.2** Binding energy per nucleon

exited state at about 2 MeV excitation. The responsibility of the tensor force for the this binding was experimentally demonstrated by observed quadrupole moment of the deuteron in the ground state. We do not have any "back of an envelope" way to short cut the calculations of the properties of the light nuclei and we switch to the nuclei above oxygen in Chap. 14.

A new picture of the nucleus emerges. The nucleus is a collection of the three quark clusters in a pion sea. Pion sharing has some features of the covalent and some of the metallic bond. When nucleons are close pions share molecular-like orbitals. Otherwise, pions are delocalised and free to move.

Quark model calculations suggest also a small additional contribution due to a *nuclear covalent bond* when two nucleons touch and their quarks share molecular-like orbitals. There is an even smaller contribution due to the colour Van der Waals interaction.

## Literature

K. Heyde, *Basic Ideas and Concepts in Nuclear Physcis* (Institute of Physics Publishing, Bristol and Philadelphia, 1999)
B. Povh et al., *Particles and Nuclei* (Springer, Berlin Heidelberg New York, 2004)

# Chapter 14
# Nuclei – Droplets of a Fermi Liquid

*Nullum est iam dictum, quod non sit dictum prius.*

Terence

It is justified to call nuclei droplets of a Fermi liquid. The nuclear force is weak; in the case of deuteron, it produces a barely bound state. In nuclei, the nucleons move almost independently of each other. In the ground State, the nucleus is, in a thermodynamic sense, at temperature $T = 0 \dim K$. As previously mentioned for liquid $^3$He, at low temperatures, a Fermi liquid may be approximated by a Fermi gas. This is also true for nuclei, where the momentum distribution only differs from gases by a smeared Fermi surface. Already in the 1930s, Fermi described the nucleus as a quantum gas using the then usual semiclassical approximation. This approximation is sufficient to understand many of the global properties of nuclei. To study the individual properties, one must, however, take into account that the nucleons move in a more-or-less spherically symmetric potential. The form of the potential may well be theoretically derived from the mean field or Hartree–Fock model. The result shows that the depth of the potential is proportional to the nuclear density. For heavy nuclei, the potential has the form of a potential well with smoothed off edges; this is called the Woods–Saxon potential:

$$V = \frac{-V_0}{1 + e^{(r-R)/a}} \, . \tag{14.1}$$

Here, $V_0$ is the depth of potential at the centre and $R$ is the nuclear radius. Individual properties of nuclei, such as the binding energies and excited states, depend on the nature of the potential (shell model). The Fermi gas approximation is good because the mean free path is large compared with the internucleon separation.

© Springer-Verlag GmbH Germany 2017
B. Povh and M. Rosina, *Scattering and Structures*,
Graduate Texts in Physics, DOI 10.1007/978-3-662-54515-7_14

*Mean field* naively means the following: the nucleon scatters off one nucleon at a time; however, there is no phase shift! In the region of the scattering centre, the wave function is modified and assumes the form of a bound state. At large separations, though, scattering has no effect on the wave function, and the nucleon behaves like a free particle. In other words, the mechanism for the long mean free path is the Pauli principle, which, in a filled Fermi sea, prevents the nucleon from choosing any other final state than its initial one.

## 14.1   Global Properties – The Fermi-Gas Model

The mean field potential that every nucleon is exposed to is produced by the superposition of the potentials of all the other nucleons and has the form of (14.1). In the nuclear volume, $V$, there are two gases, of neutrons and of protons. Because each orbital state can be occupied by two fermions of the same sort, $N$ neutrons and $Z$ protons can be contained in a nucleus, where

$$N = 2 \cdot \frac{4\pi (p_F^n)^3 \mathcal{V}}{3(2\pi\hbar)^3} \quad \text{and} \quad Z = 2 \cdot \frac{4\pi (p_F^p)^3 \mathcal{V}}{3(2\pi\hbar)^3} \,. \tag{14.2}$$

Setting the nuclear volume to

$$\mathcal{V} = \frac{4\pi}{3} R^3 = \frac{4\pi}{3} R_0^3 A \tag{14.3}$$

and using the value of $R_0 = 1.21 \dim fm$ found from electron scattering, one obtains the following Fermi momentum for a nucleus with $N = Z = A/2$ and the same radius for the proton and neutron potential wells:

$$p_F = p_F^n = p_F^p = \frac{\hbar}{R_0} \left( \frac{9\pi}{8} \right)^{1/3} \approx 250 \,\text{MeV}/c \,. \tag{14.4}$$

This is not very surprising because each nucleon has the volume of a sphere with radius $R_0$ available to it, and one can thus expect that $R_0 \cdot p_F \approx \hbar$ so $R_0 \approx \lambda_N$. This expectation agrees well with (14.4). This is, furthermore, a confirmation that the coarse estimate of the average separation between the constituents is comparable with that of the de Broglie wavelength; this is the case for all degenerate systems. The energy of the highest occupied state, the Fermi energy, $E_F$, is

$$E_F = \frac{p_F^2}{2M} \approx 33 \,\text{MeV} \,, \tag{14.5}$$

where $M$ is the nucleon mass. The typical binding energy per nucleon is $-8\,\text{MeV}$. The Coulomb repulsion and the surface energy diminish the binding energy by $8\,\text{MeV}$ per nucleon. This means that the pure nuclear binding is $B' = -16\,\text{MeV}$ per nucleon. The resulting potential depth is thus $-V_0 = B' - E_F \approx -50\,\text{MeV}$.

## 14.2 Individual Properties – Shell Model

In the previous section, we considered a potential well with many nucleons and took the nucleus to be an almost macroscopic droplet of nuclear matter. In such a model, the nuclear interaction can be used to approximately calculate the density of nuclear matter and the binding energy per nucleon, as well as the surface, Coulomb and pairing energies, but not to extract individual properties. We will again take a Fermi gas (a model of independent particles) as our approximation, but now in a spherically symmetric potential well (Fig. 14.1). Due to this spherical symmetry, the angular momenta, rather than the linear momenta, of the one-particle states are good quantum numbers, and one calculates – as in the atomic shell model – with spherical waves rather than plane waves. Because of the degeneracy of such states, the single particle states are grouped into shells. The nuclear potential is not like a Coulomb potential but more closely resembles a potential well, or, for light nuclei, a harmonic oscillator and so the magic numbers of the closed shells are not the same as in the noble gases. Experimentally, one finds especially strongly bound nuclei with large excitation energies and high separation energies for the proton or neutron numbers: 2, 8, 20, 28, 50, 82 and 126. The first three correspond to the harmonic oscillator (2, 8, 20, 40, 70, 112). The rest indicate a very strong $(\ell \cdot s)$ coupling. This is why the levels $f_{7/2}$, $g_{9/2}$, $h_{11/2}$, $i_{13/2}$ join the lower energy shells and thus the magic numbers are increased by $2 \cdot (2j + 1)$ – compared with the oscillator potential.

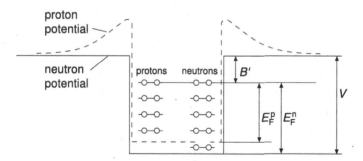

**Fig. 14.1** Sketch of the shape of the potential and proton and neutron states in the Fermi gas model

In a complete mean field calculation, there is a spin-orbit potential, as well as the central potential. In the shell model one thus uses the following effective nuclear potential.

$$V(r) = V_{\text{centr}}(r) + V_{ls}(r)\frac{\langle \ell \cdot \mathbf{s} \rangle}{\hbar}. \qquad (14.6)$$

The individual properties can be particularly well seen in nuclei where one nucleon number is magic and the other differs from a magic number by one (Fig. 14.2). Many properties, e.g., certain excitation energies, magnetic moments and matrix elements for electromagnetic and weak transitions, mostly depend only on the valence nucleon or hole.

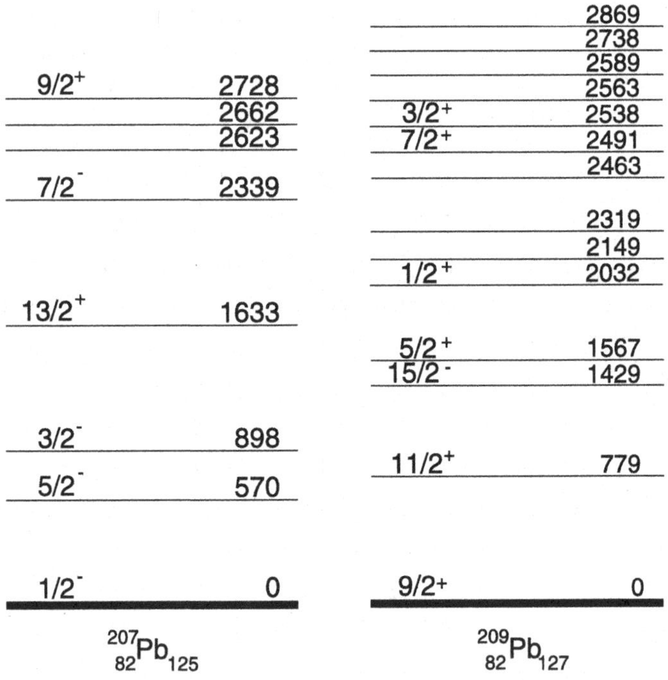

**Fig. 14.2** The single-particle states of the shell model are easily recognisable in the excitation spectrum of lead isotopes. The lead isotope $^{208}_{82}\text{Pb}_{126}$ with 82 protons and 126 neutrons has closed neutron and proton shells. The neutron hole in $^{207}_{82}\text{Pb}_{125}$ corresponds to the levels of the last closed shell: $3p_{1/2}, 2f_{5/2}, 3p_{3/2}, 1i_{13/2}, 2f_{7/2}, 1h_{9/2}$. The neutron in $^{209}_{82}\text{Pb}_{127}$ occupies one of the levels of the $2g_{9/2}, 1i_{11/2}, 1j_{15/2}, 3d_{5/2}, 4s_{1/2}, 2g_{7/2}, 3d_{3/2}$ valence shells. The levels for which spin is not given correspond to more complicated configurations. The levels next to each other are not drawn to scale. The energies are given in keV

In heavy nuclei, also the holes in the low-lying shells can be described in terms of single-particle properties; but the energy of such a state is very broad due to its short lifetime. To investigate low-lying states, it is convenient to use a hyperon $\Lambda$ as a probe because it is not subject to the Pauli principle with respect to nucleons and thus it descends alone in a cascade from the uppermost to the lowest level. Such experiments have been performed for light hypernuclei (nuclei with a hyperon), while for heavy nuclei, $\Lambda$ has only been measured in higher states.

## 14.3  Collective Excitations

### 14.3.1  Vibrational States

The most characteristic collective vibrational excitations are the giant dipole resonance and surface oscillations. These excitations may be particularly clearly demonstrated in measurements of electromagnetic transition probabilities. This can only be explained by assuming that several nucleons coherently contribute to the electromagnetic transition. Both types of vibrational excitation are quite natural in a classical liquid droplet. A giant dipole resonance corresponds to opposing vibrations of the protons and neutrons and can be viewed as analogous to the plasmon excitation of an ionised plasma or the phonon excitation of the photon branch in a crystal with ionic bonding (see Chap. 8). The surface of any water droplet can be brought into oscillation. However, nuclei are quantum systems and the nature of the collective excitations is determined by the level structure of the degenerate Fermi liquid. In the following, we would like to show that the properties of the collective vibrational states can be explained from the shell structure of the single particle excitations close to the Fermi energy.

### 14.3.2  Model

In Fig. 14.3, the excitations with $J^\pi = 1^-$ and $J^\pi = 2^+$ in a spherically symmetric nucleus with a $J^\pi = 0^+$ ground state are sketched. All these states are obtained by lifting a nucleon out of the ground state. In the shell model, the lower $J^\pi = 2^+$ states are generated by a recoupling of the angular momenta, such that all the nucleons stay in the same lowest shell. The $J^\pi = 1^-$ states correspond to an excitation of the nucleons into the next shell with opposite parity. The internucleon interaction, which is not contained in the shell model potential, mixes states with the same angular momentum. For example, for the $J^\pi = 1^-$ states of the giant resonance, the so-called particle-hole excitations, in which a nucleon is found in an excited shell and a nucleon is missing in the core of the nucleus, can mix. This change in the nucleonic configuration can be simulated by means of an effective particle–hole

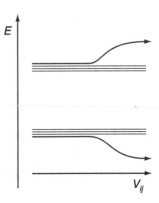

**Fig. 14.3** Level splitting in degenerate $1^-$ and $2^+$ states after a perturbation, $V_{ij}$, has been switched on. For the $1^-$ states, we have assumed a repulsive perturbation, while for the $2^+$ states, we have taken it to be attractive

interaction. This is such a strong interaction that one can view the states inside a shell as degenerate. A mixing of two degenerate states by a further interaction leads to a symmetric splitting of both states. For $N$ degenerate states, though, only one state is split from the rest when the matrix elements of the interaction have the same phase. This state – the collective state – displays a coherent superposition of the $N$ states. Let us denote the Hamiltonian operator for a nucleon in the nuclear potential by $H_0$ and the particle–hole interaction by $V$. The unperturbed particle–hole states $|\psi_i\rangle$ are the solutions of $H_0$,

$$H_0|\psi_i\rangle = E_i|\psi_i\rangle . \tag{14.7}$$

The solution $|\Psi\rangle$ of the Schrödinger equation for the total Hamiltonian is found from the relation

$$H|\Psi\rangle = (H_0 + V)|\Psi\rangle = E|\Psi\rangle . \tag{14.8}$$

The wave function, $|\Psi\rangle$, projected onto the space spanned by the states $|\psi_i\rangle$ may be written as

$$|\Psi\rangle = \sum_{i=1}^{N} c_i|\psi_i\rangle . \tag{14.9}$$

The coefficients, $c_i$, satisfy the secular equation

$$\begin{pmatrix} E_1 + V_{11} & V_{12} & V_{13} & \cdots \\ V_{21} & E_2 + V_{22} & V_{23} & \cdots \\ V_{31} & V_{32} & E_3 + V_{33} & \cdots \\ \vdots & \vdots & \vdots & \ddots \end{pmatrix} \cdot \begin{pmatrix} c_1 \\ c_2 \\ c_3 \\ \vdots \end{pmatrix} = E \cdot \begin{pmatrix} c_1 \\ c_2 \\ c_3 \\ \vdots \end{pmatrix} . \tag{14.10}$$

For simplicity, we assume that all the $V_{ij}$'s are the same,

$$\langle\psi_i|V|\psi_j\rangle = V_{ij} = V_0 . \tag{14.11}$$

The solution of the secular equation yields, for the coefficients,

$$c_i = \frac{V_0}{E - E_i} \sum_{j=1}^{N} c_j , \tag{14.12}$$

where $\sum_j c_j$ is a constant. Summing both sides over all $N$ particle–hole states and taking into account that $\sum_i c_i = \sum_j c_j$, leads to the requirement

$$1 = \sum_{i=1}^{N} \frac{V_0}{E - E_i} , \tag{14.13}$$

on the solution of the secular equation. The solutions of this equation are best expressed graphically (Fig. 14.4).

The right-hand side of (14.13) has poles at the points $E = E_i$. The solutions, $E_i'$, are found where the right-hand side is equal to one. The new energies are marked on the abscissa. The $(n - 1)$ eigenvalues are trapped between the unperturbed energies, $E_i$. The outlier, denoted by $E_C$, is the collective state. For a repulsive interaction ($V_0 > 0$), the collective state lies above the particle–hole states, while for an attractive interaction ($V_0 < 0$), it is below them (Fig. 14.4). For the giant resonances, we have again employed the same formalism as used in Chap. 6 (for the pion) and Chap. 9 (for the localised oscillatory mode) so as to demonstrate the similarity of the mechanisms that generate collective states. To get a quantitative estimate of the energy shift, we assume that $E_i = E_0$ for all $i$. Then (14.13) may be written as

$$1 = \sum_{i=1}^{N} \frac{V_0}{E_C - E_0} , \tag{14.14}$$

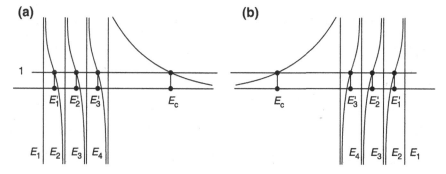

**Fig. 14.4** Graphical representation of the solution of the secular equation: (**a**) for a repulsive potential ($V_0 > 0$), (**b**) for an attractive one ($V_0 < 0$). $E_i$ are nonperturbed energies, $E_i'$ the new ones, $E_C$ is the energy of the collective state

which implies for a repulsive interaction,

$$E_C = E_0 + N \cdot V_0, \tag{14.15}$$

or in the case of an attractive interaction,

$$E_C = E_0 - N \cdot |V_0| . \tag{14.16}$$

The expansion coefficients of the collective state

$$c_i^{(C)} = \frac{V_0}{E_C - E_i} \sum_j c_j^{(C)} \tag{14.17}$$

all have the same sign and are almost independent of $i$, so long as the energy of the collective state $E_C$ lies far from $E_i$. In this approximation, the collective state can be written as

$$|\Psi_C\rangle = \frac{1}{\sqrt{N}} \sum_i |\psi_i\rangle . \tag{14.18}$$

In Fig. 14.5, the most important collective excitations are sketched. It still needs to be shown that the collective state indeed differs from the other states through its large transition probability into the ground state. The matrix element for a multipole excitation of the collective state is

$$\begin{aligned}
\mathcal{M}_C &= \int d^3x \left( c_1^{(C)} (\langle\psi_1| + c_2^{(C)} \langle\psi_2| + \ldots \right) \mathcal{O}|0\rangle \\
&= \sum c_n^{(C)} A_n \approx \frac{1}{\sqrt{N}} \sum A_n,
\end{aligned} \tag{14.19}$$

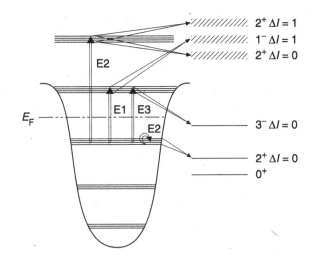

**Fig. 14.5** Collective excitations in the framework of the shell model. The collective states in which protons and neutrons oscillate in phase ($\Delta I = 0$) correspond to oscillations in the shape. They are shifted to lower energies. The collective states in which they oscillate out of phase ($\Delta I = 1$) are shifted to higher energies. E1, E2 and E3 denote the electric dipole, quadrupole and octupole excitations

where $\mathcal{O}$ is the transition operator. The integrals

$$A_n = \int \mathrm{d}^3x \, \langle \psi_n | \mathcal{O} | 0 \rangle \tag{14.20}$$

represent the amplitudes for a particle–hole excitation. For constructive interference, not only the matrix elements of the interaction, which determine the coefficients $c_n^{(C)}$, but also the excitation amplitudes, $A_n$, have to be coherent, i.e., have the same phases. It is no coincidence that there are such collective excitations in nuclei; it is rather a consequence of the fact that the transition and energy operators may be expanded as multipoles in the same manner. This implies that the transition operator will be coherent when the energy operator is coherent.

### 14.3.3 Deformation and Rotational States

In contrast with atoms, which are spherically symmetric, the majority of nuclei are deformed and can take on either a prolate (cigar shaped) or oblate (pancake shaped) form. Electrons in an atom repel each other and thus distribute themselves uniformly in the valence shell. Nuclei are, though, only spherically symmetric in the neighbourhood of closed shells. Single particle states in the valence shell are, to a first approximation, degenerate and the valence nucleons can distribute themselves nonuniformly. The attraction gathers nucleons either around the pole (prolate ellipsoid) or around the equator (oblate ellipsoid). The typical ratio of the axes of the ellipsoid in the ground state of heavy nuclei can be up to 1.3:1 and, in highly excited rotational states, even 2:1. The deformation can be statically observed by measuring the quadrupole moment. The deformation is especially spectacular in the rotational dynamics of nuclei (Fig. 14.6).

The levels follow the typical excitation pattern of a rotator: $E_J = \hbar^2 J(J+1)/(2\mathcal{I})$, where $\mathcal{I}$ is the moment of inertia of the nucleus. Nuclei do not rotate as rigid rotators; the moment of inertia is around one third of that of a rigid rotator. This is a clear indication that nuclei are made out of a Fermi liquid.

### 14.3.4 Deformation Versus Cooper Pairs

Consider two nucleons in the same orbitals outside a closed shell. The binding energy is maximised when their angular momenta are coupled to $J^\pi = 0^+$. For such a coupling, the probability is maximal for the nucleons to be close to each other, and thus the stronger attraction at small separations is optimised. The nucleus stays spherically symmetric. We call such coupled two-nucleon states Cooper pairs, in analogy to superconductivity. The wave function that describes these Cooper pairs

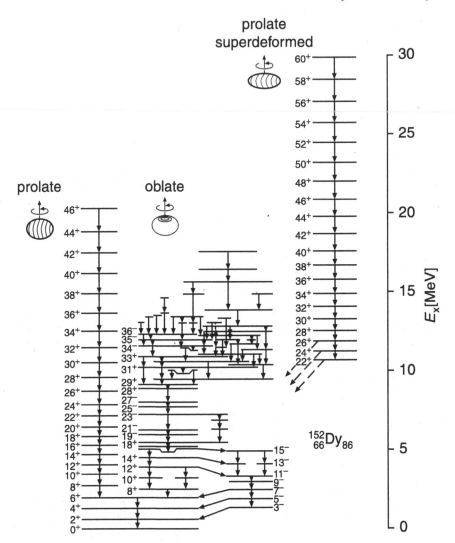

**Fig. 14.6** Energy levels in the $^{152}$Dy nucleus (from Shapey–Schafer). While the low-energy states display somewhat unusual vibrational bands, for high excitations, rotational bands are formed, which imply that the nucleus is highly deformed

is a superposition over all pairs of magnetic quantum numbers $(m_1, m_2)$, such that $m_2 = -m_1$.

For several nucleons, there is a competition between pairing and deformation. For Cooper pairs, the magnetic quantum numbers of single-particle states are uniformly occupied, while, in the deformed wave function, only the highest (or only the lowest) $|m|$-values occur. In a state with several Cooper pairs, the nucleons are only pairwise correlated and the binding energy per pair is roughly constant. In a deformed state,

**Fig. 14.7** The nuclear energy as a function of deformation for (**a**) closed shells, (**b**) a few valence nucleons, and (**c**) many valence nucleons. The deformation, $\delta = \Delta R / R$, is defined as the ratio between the difference of the larger and smaller semiaxes of the ellipsoid $\Delta R$ and the radius of a sphere with the same volume as the ellipsoid

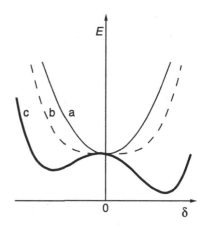

however, all the valence nucleons are correlated with each other. For a small number of nucleons, pairing dominates, while, for a larger number, deformation prevails: this is because the pairing energy grows linearly with the number of valence nucleons, but the increase in the deformation energy is quadratic.

The transition from a nucleus with closed shells to a strongly deformed nucleus is shown in Fig. 14.7. The energy as a function of the deformation has, for a nucleus with closed shells, a very steep minimum, and the frequency (energy) of quadrupole oscillations is very high. In nuclei with nearly closed shells (few valence nucleons), pairing and deformation are equally strong, which implies an almost perfect balance. Thus, the frequency of the quadrupole oscillation is very small and the energy of the first $2^+$ state is only 0.5 MeV; the vibrational spectrum is clearly visible. For a larger number of valence nucleons, deformation dominates, the round shape becomes unstable and the symmetry is spontaneously broken. Although the interaction has a spherical symmetry the lowest energy state is not spherically symmetric.

## Literature

H. Frauenfelder, E. Henlez, *Subatomic Physics* (Prentice-Hall, Englewood Cliffs, 1991)

K. Heyde, *Basic Ideas and Concepts in Nuclear Physcis* (Institute of Physics Publishing, Bristol, 1999)

B. Povh et al., *Particles and Nuclei* (Springer, Berlin, 2004)

Shapey-Schafer, Phys. World **3**(9), 31 (1990)

# Chapter 15
# Stars, Planets, and Asteroids

*Verdoppelt sich der Sterne Schein,*
*Das All wird ewig finster sein.*

Goethe

Nuclear reactions play an important role in the life of stars. For the major part of their lives, stars may be viewed as fusion reactors: nuclear reactions supply the energy needed to keep the temperature of the star constant, while gravitation ensures that the plasma is confined. The final stages of stars are understood in terms of degenerate fermion systems.

Gravitation is the dominant force in systems of astronomical dimensions. Our experience, primarily based on the mechanics of the solar system and the motion of terrestrial satellites, has taught us that the nature of gravitational systems is ruled by the virial theorem. This also holds for the largest stars, our sun and other stars in the main sequence, white dwarfs, neutron stars, the planets and asteroids. We wish to show that the properties of these objects may be qualitatively understood from atomic constants and the virial theorem.

## 15.1 The Sun and Sun-Like Stars

The stars of the main sequence are produced by the contraction of interstellar gas and dust. This material is almost totally composed of primordial hydrogen and helium, which were produced in the big bang, plus around 2% of heavier elements. The contraction heats up the centre of the star. When the temperature and pressure are large enough to make the fusion of nuclei possible, the star enters thermal equilibrium. The star then ceases its contraction and the energy that is radiated away is compensated by energy production at the centre of the star. The energy produced in nuclear reactions is primarily transported to the surface by radiation. This does not significantly mix

© Springer-Verlag GmbH Germany 2017
B. Povh and M. Rosina, *Scattering and Structures*,
Graduate Texts in Physics, DOI 10.1007/978-3-662-54515-7_15

the stellar material. During the star's lifetime, its chemical composition is naturally changed in the regions where nuclear reactions take place, thus, most of all, in the centre of the star.

### 15.1.1 Equation of State

The pressure, $p$, at a radius $r$ in a star may be calculated assuming hydrostatic thermal equilibrium, i.e., that the gravitational force, $F_g$, produced by gravitational pressure at a radius $r$ (Fig. 15.1)

$$dF_g = -\frac{GM_r dm}{r^2} = -\frac{GM_r \rho}{r^2} dAdr, \qquad (15.1)$$

must be balanced by the force $dF_p = -dpdA$, produced by thermal pressure. In (15.1), $G$ is the gravitational constant, $\rho$ the density at position $r$ and $M_r$ is the mass that is contained inside a sphere of radius $r$,

$$M_r = \int_0^r 4\pi \rho r^2 dr. \qquad (15.2)$$

Equilibrium implies $dF_g + dF_p = 0$, which leads to the following equation of state for the condition of hydrostatic equilibrium:

$$\frac{dp}{dr} = -\frac{GM_r \rho}{r^2}. \qquad (15.3)$$

This equation, significantly refined by taking the chemical composition of the star and other details into account, has been studied for all possible scenarios. To find a qualitative understanding of stellar behaviour, we will take the density, $\rho$, to be constant. Then we may replace the differential quantities, $dr$ and $dp$, by the integrated

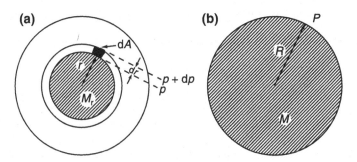

**Fig. 15.1** A realistic equation of state must take into account that the gravitational force and the force of thermal pressure at radius $r$ have to be in equilibrium

variables: the stellar radius $R$ and the confining pressure $P$. In this way, (15.3) reduces to

$$\frac{P}{R} = -\frac{GM}{R^2}\frac{M}{V}. \tag{15.4}$$

This simplification is thoroughly acceptable for cold objects. The density of white dwarfs and neutron stars does not depend much on the radius. Hot stars, though, have a massive core, where the majority of the total mass is concentrated, and we cannot simply insert the stellar radius into (15.4). Instead, we rather use the average separation, $d$, between the plasma's constituents. This scale sets the thermodynamic properties of a star. In Fig. 15.1, we sketch the transition from a realistic ansatz (15.3) of the equation of state (Fig. 15.1a) to a simplified stellar model (15.4), with constant density (Fig. 15.1b).

## 15.1.2 Virial Theorem

Consider a star with mass $M$, radius $R$ and constant density $\rho = M/V$. The star's potential energy is then

$$E_{\text{pot}} = -\frac{3}{5}\frac{GM^2}{R}. \tag{15.5}$$

The total energy of the star is given by the sum of the kinetic (here, we prefer to call it thermal) energy and the potential energy,

$$E = E_{\text{therm}} + E_{\text{pot}}. \tag{15.6}$$

For the nonrelativistic case, the following relation holds:

$$E = \frac{1}{2}E_{\text{pot}} = -E_{\text{therm}}. \tag{15.7}$$

This is the well-known form of the virial theorem for a $1/r$-potential. The star exists stably at the minimum of the total energy.

## 15.1.3 Size and Temperature

In the following estimate, we consider a star built solely from hydrogen. Instead of $G$, we use – in analogy to the fine structure constant, $\alpha$ – the dimensionless coupling constant, $\alpha_G$,

$$\alpha_G = \frac{Gm_{\text{p}}^2}{\hbar c} \approx 10^{-38}. \tag{15.8}$$

Here, $m_p$ is the proton mass. It is also useful to express the stellar mass in terms of the number of nucleons, $N$. We write $M = N(m_p + m_e) \approx N m_p$. The potential energy is then

$$E_{pot} = -\frac{3}{5} \frac{\alpha_G \hbar c N^2}{R}.$$ (15.9)

Let us consider stars in which radiation pressure is small compared with non-relativistic particle pressure. This holds for the sun, somewhat more massive stars and, especially, for objects smaller than the sun. In such objects, the gravitational pressure is balanced by the pressure due to the thermal motion of the $N$ protons and $N$ electrons.

### 15.1.4  Proton Energy

The average kinetic energy of a protons or an electron is $(3/2)kT$. The total kinetic energy of the star is thus

$$3NkT = -\frac{1}{2}E_{pot} = \frac{1}{2}\frac{3}{5}\frac{\alpha_G \hbar c N^2}{R}.$$ (15.10)

If we now replace the radius by the average separation between the protons, $d$, i.e., $R^3 \approx Nd^3$, then the relation between the temperature, the average separation, $d$, and the number of particles in the star, $N$, is

$$3kT = \frac{3}{10}\frac{\alpha_G \hbar c N^{2/3}}{d}.$$ (15.11)

If we denote the number of nucleons in the sun, $10^{57}$, by $N_0$, then, by a whim of nature, $\alpha_G = N_0^{-2/3}$! In astronomy, it is common to normalise masses via the solar mass and write the relation between the average particle separation, the mass and radius as

$$kT = \frac{1}{10}\left(\frac{N}{N_0}\right)^{2/3}\frac{\hbar c}{d}.$$ (15.12)

### 15.1.5  Electron Energy

As stars contract, the average separation between the electrons keeps shrinking and when $d$ becomes comparable with the electron de Broglie wavelength, the degeneracy pressure of the electrons becomes more important. The average kinetic energy of an electron in a degenerate electron gas can be estimated to be $\hbar^2/2m_e d^2$. A more exact calculation based on (9.8) yields an additional factor, $\frac{3}{5}(9\pi/4)^{2/3} \approx 2.2$. (Note that

**Fig. 15.2** Dependence of
stellar temperature on the
average separation, $d$,
between the protons

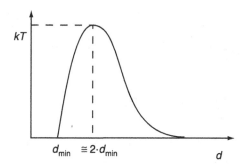

the definition of $d$ used in this chapter differs slightly from (9.8).) This yields our
simplified condition for a star to be in hydrostatic equilibrium,

$$\frac{3}{2}kT + 2.2 \frac{\hbar^2}{2m_e d^2} = \frac{3}{10}\left(\frac{N}{N_0}\right)^{2/3} \frac{\hbar c}{d}. \tag{15.13}$$

At large separations, $d$, the second term becomes insignificant and the electron energy
may be treated classically. The electron energy, as the consequence of the equiparti-
tion theorem, is equal to the proton energy and also contributes $\frac{3}{2}kT$. The dependence
of the temperature on $d$ is sketched in Fig. 15.2.

## 15.1.6 White Dwarfs

Consider a star, which, at the end of its lifetime, is a white dwarf with a solar mass.
We will neglect the fact that, in the last stage of its life as a "small" red giant, it loses
some mass. From (15.13), the star's contraction is stopped by electron pressure at
$d_{min} \approx 3.5\lambda_e$. This implies that the radius of the white dwarf is $10^4$ km. The maximal
temperature is reached at $d = 2d_{min}$; it is

$$kT \approx \frac{1}{70}\left(\frac{N}{N_0}\right)^{2/3} m_e c^2 \approx 7\,\mathrm{keV} \approx k \cdot 10^8\,\mathrm{K}. \tag{15.14}$$

This is a pretty good estimate of the temperature of the core of a red giant! At this
temperature, the energy of the star is won by burning helium. Anyway, hydrogen
burning takes place already at $kT \approx 1\,\mathrm{keV}$. (In the centre of the sun, $kT = 1.3\,\mathrm{keV}$.)
The life of a star can be described as follows (Fig. 15.2): the star contracts until it
reaches the temperature $kT \approx 1\,\mathrm{keV}$. It keeps this temperature until the hydrogen
in the star's core is used up. The star's core then contracts until it reaches the temperature $kT_{max} \approx 10\,\mathrm{keV}$, while the stellar mantle expands and the surface cools to
a temperature around 3000 K, so that the star appears red. After the helium in the
core has been used up, essentially it is fused into carbon and oxygen, the mass of

the star is too small to let further contraction generate a high enough temperature to ignite further nuclear reactions. The stellar core then cools to a white dwarf and the gravitational pressure is compensated by the Pauli pressure due to the electrons.

### 15.1.7  Brown Dwarfs

According to (15.14), stellar objects with only a few hundredths of the solar mass only reach $kT_{max} \leq 1\,\text{keV}$. This temperature is too low to win further energy from nuclear reactions. The life of such stellar objects is very simple. They contract and the particles' kinetic energy increases. However, this increase of kinetic energy is only half as large as the decrease in potential energy. The difference is radiated off. Due to their low surface temperatures, such dwarfs only shine very feebly, most brightly at the time when they are around their maximal temperature. The colour of "brown" dwarfs is actually reddish, but the name "red dwarf" is reserved for normal stars with masses between 0.1 and 1 $M_\odot$ solar masses.

## 15.2  Energy Production in the Sun

The temperatures, which we have calculated, yield the maxima of the Maxwell distributions. Due to the repulsion of charged nuclei, only those in the high energy tail of the distribution (Fig. 15.3) fuse together. The cross-section in the case of Coulomb repulsion is proportional to the probability that the reaction participants enter the interaction range, the so-called Gamow factor, $\exp(-b/E^{1/2})$, and to the

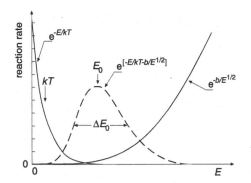

**Fig. 15.3** Sketch of the product of the Maxwell distribution $\exp(-E/kT)$ and the Gamow factor $\exp(-b/E^{1/2})$ as required to calculate the rate of fusion reactions. The product of these curves is proportional to the fusion probability (*dashed curves*). Fusion essentially takes place in a rather narrow energy interval around $E_0$ of width $\Delta E_0$. The integral over this curve is proportional to the total reaction rate

strength of the interaction between the nuclei. Here, the factor $b = \pi \alpha Z_1 Z_2 \sqrt{2mc^2}$, where $m$ is the reduced mass and $Z_1$ and $Z_2$ are the charges of the fusing nuclei. Fusion can take place via the strong interaction through particle emission, via the electromagnetic through gamma emission or via the weak force through electron neutrino emission.

The Gamow factor is one of the nicest examples of tunnelling in quantum mechanics. It is known from basic quantum mechanics that the probability of tunnelling effects decreases as $e^{-2G}$, where, for a rectangular barrier of height $V_0$ and length $L$, the exponent is $2G = 2\sqrt{2m(V_0 - E)}\,L/\hbar$. The exponents in the Gamow factor can be understood such that the Coulomb potential has an effective height $\overline{V - E} = (\pi/2)^2 E$ and an effective range $L = \alpha Z_1 Z_2 \hbar c/E$. (This is the length of the barrier up to the turning point radius, $L$, where $V - E$ vanishes, $\alpha Z_1 Z_2 \hbar c/L = E$).

The reaction rates of fusion reactions per unit volume, $W$, via the strong interaction are given by the following relation:

$$W = n_1 \cdot n_2 \cdot \langle \sigma v \rangle, \tag{15.15}$$

where $n_1$ and $n_2$ are the number densities of the nuclei taking part in the fusion and $< \sigma v >$ is the averaged product of the fusion cross-section (Fig. 15.3) and the relative velocity of the fusion partners.

We are, though, especially interested in two reactions that do not follow this pattern. The lifetime of the sun, which is around $10^{10}$ years, is determined by the weak process $p + p \rightarrow d + e^+ + \nu_e$. Primordially, only hydrogen and helium (and tiny amounts of some light elements) were available. The time scale of the construction of the heavy elements is determined by $3\alpha \rightarrow {}^{12}C$ process in red giants.

## 15.2.1 Proton–Proton Cycle

According to the solar model, nuclear fusion takes place in a core of 70% of the solar mass and with radius $R \approx R_\odot/3$. The temperature at the centre of the sun reaches $kT \approx 1.3 \text{keV}$ and falls off sharply with the distance from the centre. To estimate the reaction rate using the average values of the temperature and density, it is better only to consider the most central core of the sun, with $R \approx R_\odot/10$. A tenth of the total solar mass is inside this volume, and the production rate can be calculated at an average energy of $kT \approx 1.0 \text{keV}$. This core is responsible for half the total energy production. The luminosity of the sun is $L = 4 \cdot 10^{26}$ W. When the sun was young, its composition of elements was that of the modern solar surface (71% hydrogen, 27% helium and 2% heavier elements). The mass of the sun, as we previously mentioned, is $M_\odot = 10^{57}$ nucleon masses, the radius is $R_\odot \approx 7 \cdot 10^8$ m. This implies that the number of protons in the central core of the sun is $N_H = 0.7 \cdot 10^{56}$ and the density is $n_H = 0.5 \cdot 10^{32}$ protons/m$^3$.

About 98% of the burning is through the proton–proton cycle, the main branch of which is

$$p + p \rightarrow d + e^+ + \nu_e + 0.42 \text{ MeV} \qquad \tau(p) = 10^{10} \text{ a}$$
$$p + d \rightarrow {}^3\text{He} + \gamma + 5.49 \text{ MeV} \qquad \tau(d) = 1.6 \text{ s}$$
$${}^3\text{He} + {}^3\text{He} \rightarrow p + p + \alpha + 12.86 \text{ MeV} \qquad \tau({}^3\text{He}) = 10^6 \text{ a}$$
$$e^+ + e^- \rightarrow 2\gamma + 1.02 \text{ MeV}.$$

All in all, during the net reaction, $4p \rightarrow \alpha + 2e^+ + 2\nu_e$, an energy of $E_{pp} = 26.72 \text{ MeV}$ is released. The first step is the slowest; it proceeds via the weak interaction. It determines the lifetime of the sun. We will only qualitatively discuss the lifetime of deuterons, which is determined by the electromagnetic interaction and the lifetime of $^3$He, which decays through the nuclear interactions.

The proton lifetime can be extracted from the luminosity and proton number in the centre of the sun,

$$\frac{L}{2} = N_H \frac{(E_{pp}/4)}{\tau(p)}. \tag{15.16}$$

This yields the proton lifetime in the sun, $\tau(p) \approx 10^{10}$ years. The solar lifetime is of the same order of magnitude.

The peculiarity of the proton–proton cycle is best represented graphically. In Fig. 15.4, we sketch the most important data for pp-fusion. $\beta$-decay from $^2$H takes place from the scattering state. Due to the long tail of the deuterium wave function, the overlap with the scattering state primarily takes place outside the range of the strong interaction below the Coulomb barrier. To calculate the reaction rate, we would have had to work out the decay probabilities for the individual energies and then integrate over them. This cannot be done analytically. To estimate it *on the back of an envelope*, we will work from the start with averaged quantities.

Let us try to estimate the proton lifetime, $\tau(p)$, with the help of the above-listed parameters of the sun. Let us first calculate the fraction of the protons that participate in fusion reactions in the proton–proton cycle at temperature $T$. The reaction rate depends on the product of the Maxwell distribution, $\propto \exp(-E/kT)$, with the Gamow factor. The product is maximal at

$$\frac{d}{dE}(E/kT + \pi\alpha\sqrt{2mc^2/E}) = 1/kT - \frac{\pi}{2}\alpha\sqrt{2mc^2}E^{-3/2} = 0, \tag{15.17}$$

where we have set $Z_1 = Z_2 = 1$ for protons. This implies at, $kT = 1.0$ keV,

$$\frac{E_0}{kT} = \frac{\pi}{2}\alpha\sqrt{\frac{2mc^2}{E_0}} \approx 5. \tag{15.18}$$

At this energy, the product of the Maxwell factor and the Gamow factor is $\exp(-5 - 10) = 3.1 \cdot 10^{-7}$.

For the effective energy interval, $\Delta E_0$, we take the distance from the maximum to the point where the reaction rate falls to a $1/e$ value (the exponent is increased by 1),

**Fig. 15.4** Shown schematically are a few of the ingredients that determine the rate of the pp-cycle. (**a**) The nuclear potential for the deuteron and the pp system. The energy scale is in MeV and the 0 energy is at the proton–neutron threshold. Because of the proton–neutron mass difference, the energy of a proton–proton system at rest is −1.3 MeV. (**b**) The wave function of the deuteron in the ground state. (**c**) The scattering wave function of the two protons. (**d**) The overlap of the two determines the transition matrix element. The main contribution is outside the $R_d$ and below the Coulomb barrier

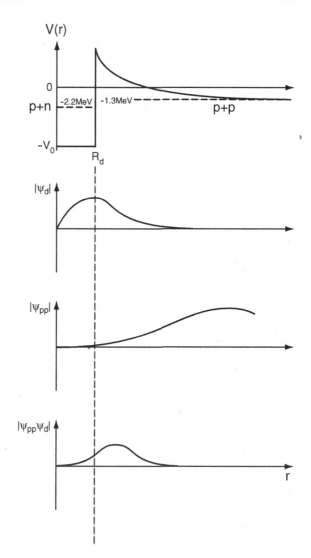

$$\frac{d^2}{dE^2}(E/kT + \pi\alpha\sqrt{2mc^2/E})\frac{(\Delta E_0)^2}{2}$$

$$= \frac{3\pi}{8}\alpha\sqrt{2mc^2}E_0^{-5/2}\Delta E_0^2 \qquad (15.19)$$

$$= \frac{3}{4}\frac{E_0}{kT}\left(\frac{\Delta E_0}{E_0}\right)^2 = 1.$$

One finds $\Delta E_0/E_0 = 0.52$.

The fraction of the protons that, at a temperature $kT = 1\,\text{keV}$, penetrate the Coulomb barrier and, for a short time, $\Delta t$ form an unstable $^2\text{He}$ nucleus is

$$B = (\Delta E_0/kT)\exp(-E_0/kT - \pi\alpha\sqrt{2mc^2/E_0}) = 8.0 \cdot 10^{-7}. \tag{15.20}$$

If we choose the deuteron radius to be $R_d \approx 4\,\text{fm}$, then the probability of a two-proton cluster inside such a separation is

$$w_{pp} = Bn_H 4\pi R_d^3/3 = 1.1 \cdot 10^{-17}. \tag{15.21}$$

For the process pp $\rightarrow$ d via the weak interaction $H_\beta$, we may apply perturbation theory (the golden rule),

$$\sigma v = \frac{2\pi}{\hbar}|M|^2\rho(E_0), \qquad M = \langle d|H_\beta|pp\rangle, \tag{15.22}$$

where $v$ is the projectile speed and $\rho(E)$ is the final-state density. The asymptotic proton density is normalised to 1 ($\psi_p(r \rightarrow \infty) \rightarrow 1$). To avoid having to evaluate the nuclear matrix element, $M$, we can estimate it from a similar process. The beta decay, $^{18}\text{Ne} \rightarrow {}^{18}\text{F}$, with the $0^+(T = 1) \rightarrow 1^+(T = 0)$ transition and $\tau_{18} \equiv \tau({}^{18}\text{Ne} \rightarrow {}^{18}\text{F}) = 2.4\,\text{s}$ (i.e., with half life 1.7 s) is appropriate for this purpose.

$$\frac{1}{\tau_{18}} = \frac{2\pi}{\hbar}|M_{18}|^2\rho(E_{18}), \qquad M_{18} = \langle {}^{18}\text{Ne}|H_\beta|{}^{18}\text{F}\rangle. \tag{15.23}$$

In both cases, p $\rightarrow$ n, and we view the other nucleons as spectators. The wave functions inside the nuclear volume,

$$\psi_p({}^{18}\text{F}) \approx \psi_n({}^{18}\text{Ne}) \approx \psi_n(d) \approx (4\pi R_d/3)^{-1/2}, \tag{15.24}$$

are of the same order of magnitude as are the state densities, $\rho(E_0) \approx \rho(E_{18})$. Only the wave function of the incoming proton is different: it corresponds to an unbound state and is diminished by the Gamow factor,

$$\psi_p(pp) \approx e^{-G} \equiv e^{(\pi\alpha/2)\sqrt{2mc^2/E_0}}. \tag{15.25}$$

The ratio between the matrix elements is thus

$$\frac{M^2}{M_{18}^2} = \frac{e^{-2G}}{(4\pi R_d/3)^{-1}}, \tag{15.26}$$

which gives us an estimate for the cross-section,

$$\sigma v \approx \frac{e^{-2G}4\pi R_d^3/3}{\tau({}^{18}\text{Ne} \rightarrow {}^{18}\text{F})}. \tag{15.27}$$

The proton lifetime, $\tau(p)$, is inversely proportional to the reaction rate with all effective protons in its vicinity, $N_{eff} = n_H(\Delta E_0/kT)\exp(-E_0/kT)$, and one obtains (with the help of the relations (15.20) and (15.21))

$$\frac{1}{\tau(p)} = N_{eff}\,\sigma\,v \approx \frac{w_{pp}}{\tau(^{18}Ne \to {}^{18}F)} = \frac{1}{2.2 \cdot 10^{17}\,s} = \frac{1}{7.2 \cdot 10^9\,a}. \qquad (15.28)$$

The reaction rate is thus the product of the probability of two-proton clusters and the beta decay rate during the contact. The proton lifetime, $\tau(p) = 7 \cdot 10^9$ years, is in very good agreement with the proton lifetime in the sun, $\tau(p) \approx 10^{10}$ years, calculated from the luminosity.

We wish to note that, in estimating the probability of the two-proton clusters, a quantum mechanical derivation was needed. Classically, next to no protons could cross the Coulomb barrier. Quantum mechanically, the protons can do it through tunnelling; the tails of the wave function at $r \sim 0$ are rather small but they are sufficient. Because there are no resonances in the pp-system (nonresonant process), the protons in the nuclear potential cannot pass backward and forward several times, but rather they recoil immediately and only a very small fraction experience the beta decay. In the next section, $(3\alpha \to {}^{12}$ C); on the other hand, we will encounter a resonant process, for which classical statistical mechanics is applicable.

The lifetime of deuterium is much shorter. The probability that deuterium and a proton enter the interaction range is, apart from the small difference in the masses of the partners, identical to the proton–proton case. The electromagnetic transitions are, though, orders of magnitude faster. In our case, this is a magnetic dipole transition. The transition probability is, if we take $\tau(d)$ into account, $\approx 10^{15}s^{-1}$, which corresponds to a gamma width of $\approx 1$ eV. These numbers may be compared with $\approx 1/2$ s and $10^{-16}$ eV for the beta decay. The third reaction proceeds via the strong interaction. All those nuclei penetrating the Coulomb barrier interact. The lifetime is determined just by the Gamow factor and the $^3$He density.

## 15.2.2 $3\alpha \to {}^{12}C$-Process

When, after $\sim 10^{10}$ years, the solar hydrogen has been so much used up that the thermal pressure cannot compensate the gravitational pressure, the core will collapse and heat up to around $2 \cdot 10^8$ K, which corresponds to $kT \approx 17$ keV. The solar mantle will greatly expand so that, despite the higher energy production, the sun will appear red (red giant). The first reaction, which at that stage is responsible for the new equilibrium, is $3\alpha \to {}^{12}C$. Because there are no stable nuclei with $A = 5$ or $A = 8$ and because there is no other way to generate carbon and heavier elements, it is worth studying this reaction in more detail.

It is a peculiarity of the synthesis of carbon that it takes place sequentially via two resonances (Fig. 15.5).

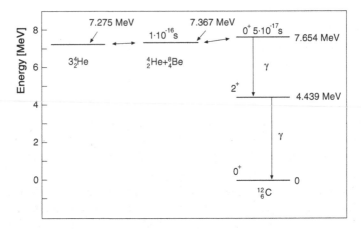

**Fig. 15.5** Energy levels of the systems: $3\alpha$, $\alpha + {}^8\text{Be}$ and ${}^{12}\text{C}$. Just above the ground states of the $3\alpha$ system and of the $\alpha + {}^8\text{Be}$ system. There is a $0^+$ state in the ${}^{12}\text{C}$ nucleus that can be created through resonant fusion of ${}^4\text{He}$ nuclei. This excited state decays with a probability of 0.04% into the ${}^{12}\text{C}$ ground state

$$\alpha + \alpha \leftrightarrow {}^8\text{Be}, \qquad {}^8\text{Be} + \alpha \leftrightarrow {}^{12}\text{C}^*, \qquad {}^{12}\text{C}^* \rightarrow {}^{12}\text{C} + 2\gamma. \tag{15.29}$$

The ${}^8\text{Be}$ ground state has a lifetime of $0.97 \cdot 10^{-16}$ s and decays into two $\alpha$'s, releasing 91.9 keV. To calculate the concentration of ${}^8\text{Be}$ in ${}^4\text{He}$ plasma, we write the equations for chemical equilibrium. The chemical potentials are the same before and after the reaction,

$$\mu_4 + \mu_4 = \mu_8 + \Delta E_8, \tag{15.30}$$

or explicitly,

$$\begin{aligned} kT \ln n_4 &\left( \frac{2\pi\hbar^2}{m_4 kT} \right)^{3/2} + kT \ln n_4 \left( \frac{2\pi\hbar^2}{m_4 kT} \right)^{3/2} \\ &= kT \ln n_8 \left( \frac{2\pi\hbar^2}{m_8 kT} \right)^{3/2} + \Delta E_8. \end{aligned} \tag{15.31}$$

Here, $n_4$ and $n_8$ are the ${}^4\text{He}$ and ${}^8\text{Be}$ densities, and $\Delta E_8 = 91.9$ keV is released in the decay. In the second step, we follow a similar procedure,

$$\mu_8 + \mu_4 = \mu_{12}^* + \Delta E_{12}^*, \tag{15.32}$$

where the asterisk signifies the excited carbon state and $\Delta E_{12}^* = 288$ keV corresponds to the 7.654 MeV excited state. The second equilibrium condition written out fully is then

$$kT \ln n_8 \left( \frac{2\pi\hbar^2}{m_8 kT} \right)^{3/2} + kT \ln n_4 \left( \frac{2\pi\hbar^2}{m_4 kT} \right)^{3/2}$$
$$= kT \ln n_{12}^* \left( \frac{2\pi\hbar^2}{m_{12}^* kT} \right)^{3/2} + \Delta E_{12}^* . \qquad (15.33)$$

One may take the helium density at the centre of a red giant to be $\rho = 10^9 \, \mathrm{kg \, m^{-3}}$ ($n_4 = 1.5 \cdot 10^{35} \mathrm{m^{-3}}$). We have assumed the average temperature $T = 10^8 \, \mathrm{K}$ ($kT = 8.62 \, \mathrm{keV}$). From this, one finds $n_8/n_4 = 6.6 \cdot 10^{-9}$ and $n_{12}^*/n_4 = 3.7 \cdot 10^{-27}$. This is very small, but it suffices!

The carbon production rate is then

$$dn_{12}/dt = n_{12}^* \Gamma_\gamma . \qquad (15.34)$$

From the experimental value of the gamma width of the 7.654 MeV state, $\Gamma_\gamma = 3.58 \, \mathrm{meV}$ ($5.6 \cdot 10^{12} \mathrm{s^{-1}}$), one obtains

$$\frac{dn_{12}/dt}{n_4} = \frac{n_{12}^*}{n_4} \Gamma_\gamma = 2.1 \cdot 10^{-14} \mathrm{s^{-1}} = \frac{1}{1.5 \cdot 10^6 \, \mathrm{a}} . \qquad (15.35)$$

This period of around a million years is a very good estimate for the duration of the helium-burning phase of a sun-like star.

Without these two resonances, carbon synthesis would be orders of magnitude slower. As with the proton–proton cycle, we would have had to estimate the collision time. However, the time for the partners being near each other is much larger in a resonant system. Otherwise, the universe would just be built out of hydrogen, helium, cosmic background radiation and, perhaps, dark matter and dark energy. Some philosophers ascribe to Nature an active role in the choice of the physical constants, choosing them in such a way that human existence is possible (the anthropic principle). This has to include the slow burning of hydrogen in the sun and also the rapid construction of the heavy elements that are necessary for life.

It is interesting to note that the excited $J^\pi = 0^+$ state in carbon at around 7 MeV was in fact predicted by Fred Hoyle (1953) on the grounds that, otherwise, the synthesis of the heavy elements would not be possible.

## 15.3 Stars More Massive than the Sun

Stars with 10 or more solar masses live quickly and intensively. Hydrogen is burnt to form $\alpha$'s through the carbon–nitrogen-oxygen (CNO) cycle, which is much quicker than the proton–proton cycle. The slowest step in the CNO cycle is $^{14}N + p \rightarrow {}^{15}O + \gamma$, which is an electromagnetic transition. In later phases of their lives, these stars continue having higher temperatures than the sun. These temperatures are so high that not just carbon but also heavier elements up to iron are produced by fusion. Neutrons, essentially generated in $(\alpha, n)$ reactions, produce the elements up to lead. When the stellar core is mostly made of iron, only endothermic nuclear reactions are possible and the star can no longer resist gravitational pressure. It implodes and then explodes.

### 15.3.1  Neutron Stars

If the mass left after the explosion – mostly made of iron – corresponds to a scale of 1.5 solar masses, then its electron Pauli pressure cannot resist the gravitational pressure and a neutron star is formed. Due to the high electron density in the imploded iron core, *inverse beta decay*

$$^{56}Fe + 26e^- \rightarrow 55n + (pe^-) + 25\nu , \tag{15.36}$$

starts up and converts almost all the protons into neutrons. Around 2% of the protons and electrons live on in dynamic equilibrium with the degenerate neutrons. The Pauli pressure in neutron stars is due to the degeneracy of the neutron states. The average separation of the neutrons is $\lambda_n$, which is a factor of 1,000 smaller than the electron separation in white dwarfs. The radii of neutron stars are of the order of 10 km, again a factor of 1,000 smaller than white dwarfs.

### 15.3.2  Black Holes

At still greater residual masses – when the Pauli pressure of the degenerate neutrons has to yield to the gravitational pressure – the star collapses still further and forms a black hole. The gravitational energy of the black hole at its surface is so large that not even photons can escape from it.

The potential energy of a photon at the surface of a star would be

$$E_{\text{pot}} = -\frac{GM}{R}\frac{\hbar\omega}{c^2} , \tag{15.37}$$

so that its kinetic energy at infinity is

$$E_{\text{kin}} = \hbar\omega' = \hbar\omega - \frac{GM}{R}\frac{\hbar\omega}{c^2} . \tag{15.38}$$

The radius of a black hole is found via $\hbar\omega' = 0$, $R \leq GM/c^2$. General relativity theory yields a value of the critical radius greater by a factor of 2. The term $2GM/c^2$ is called the Schwarzschild radius.

### 15.3.3 Element Abundance

The isotope abundance in terrestrial, lunar and meteoritic samples is, with a few exceptions, universal and agrees with the nuclide abundance in cosmic rays that originate from outside the solar system (Fig. 15.6). Our current understanding is that the synthesis of the deuterium and helium available today took place in the early stage of the universe, when it was just a few minutes old.

The elements from carbon to uranium are produced in the final stages of heavy stars. In the red giant stage, the elements carbon and oxygen are produced, while in later stages, the elements up to iron are made. Successive neutron captures produce neutron-rich isotopes. If the isotopes are $\beta$ unstable, they decay into a stable isobar. In this way, ever heavier elements are produced along a stability valley. In this slow process (s-process), the nuclei up to lead are generated. Nuclei above lead are $\alpha$ unstable and decay into $\alpha$ particles and lead. The rapid process (r-process) probably takes place during supernova explosions. In this stage, neutron fluxes of $10^{32}$ m$^{-2}$ s$^{-1}$

**Fig. 15.6** Element abundances in the solar system as a function of the mass number $A$. The silicon abundance was normalised to $10^6$

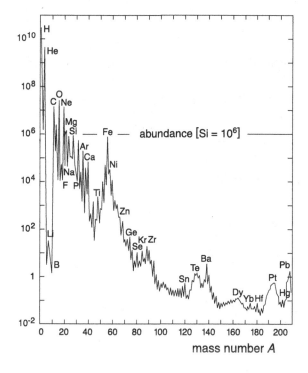

are achieved and the successive absorbtion of many neutrons is much quicker than the $\beta$ and $\alpha$ decay processes. In this fashion, elements heavier than lead are produced. The upper limit on the production of the transuranic elements is provided by spontaneous fission.

## 15.4   Planets and Asteroids

We now wish to estimate how large the masses of the largest planets are and how to draw the line between planets and asteroids. Planets and asteroids are names for objects where the average separation between protons is larger than the Bohr radius (this implies that they are made of normal solid state),

$$d \geq a_0 = \frac{\hbar c}{\alpha m_e c^2} . \tag{15.39}$$

From (15.13), we can obtain the mass of the largest planet,

$$\frac{N}{N_0} \leq \left(\frac{10\alpha}{3}\right)^{3/2} \approx 4 \cdot 10^{-3} , \tag{15.40}$$

which is a few thousandths of the solar mass. This is of the same order as the mass of Jupiter.

We set the lower limit of the mass of a planet by considering an object the radius of which is much larger than the height of its mountains. We will see that the maximal height of mountains is determined by the mass of the planet or asteroid.

The upper bound on a mountain is reached when the weight of the mountain liquefies the material of the base. Liquid signifies here that the stone becomes an amorphous substance with a very high viscosity, such as the aggregate state of the Earth's mantle, on which the Earth's crust swims.

In Fig. 15.7, the most important quantities are shown. The stability limit is given when decreasing the height of the mountain by $\Delta h$ leads to a decrease in the potential energy equal to the melting energy,

**Fig. 15.7** The weight of a mountain of height $h$ liquefies the stone and squeezes it to the sides

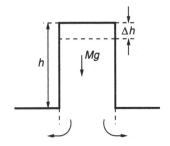

$$Mg \cdot \Delta h = E_{\text{liq}} \cdot n\Delta hX . \tag{15.41}$$

Let us denote the number of molecules per unit volume by $n$, the melting energy per molecule by $E_{\text{liq}}$, their mass number by $A$ and the surface of the base of the mountain by $X$, and $g$ is the gravitational field strength (free-fall acceleration) on the Earth. If we substitute the mass of the mountain by

$$M = nAm_{\text{p}}hX \tag{15.42}$$

into (15.41), the stability condition becomes

$$gnAm_{\text{p}}hX \leq E_{\text{liq}}nX , \tag{15.43}$$

where

$$h \leq \frac{E_{\text{liq}}}{Am_{\text{p}}g} . \tag{15.44}$$

Let us estimate $E_{\text{liq}}$ and express its size through atomic quantities: the typical binding energies for silicates, which make up the major part of the Earth's crust and the Earth's mantle, is a few eV, or, using Rydberg's constant, $\approx 0.2$ dim $Ry$. The melting energy of water is $\approx 1/8$ of the binding energy. So 10% of the binding energy is a reasonable approximation for $E_{\text{liq}}$. Thus, the condition (15.44) expressed in "fundamental" scales is

$$h \leq \frac{0.02\,\text{Ry}}{Am_{\text{p}}g} . \tag{15.45}$$

For our Earth, this estimate yields $h \leq 30\,\text{km}$. Due to erosion, the height of Mount Everest ($h \approx 10\,\text{km}$) is less relevant than the thickness of the Earth's crust, which swims on the Earth's mantle. The reason that the mantle is liquid, or, better said, a viscous liquid, is the same as that used above to estimate the maximal mountain height. The thickness of the Earth's crust is 12–62 km, which is in extremely good agreement with our estimate of 30 km.

We expect that the radius of a planet must be much larger than the height of its mountains: $h_{\text{max}}/R \leq 0.1$ is probably a good choice for the following estimate. The ratio $h_{\text{max}}/R$ for the Earth is $0.5 \cdot 10^{-2}$, if we replace the thickness of the Earth's crust (30 km) as a measure of the height. The average densities of planets only vary by a factor of 2–3, so we may take $g = GM/R^2 \propto R$ as the acceleration due to gravity at the planetary surface. From (15.45), the ratio

$$\frac{h_{\max}}{R} \propto \frac{1}{R^2} \tag{15.46}$$

follows.

Pluto as well as the moon fulfill this criterion for being a planet. The largest asteroid, Ceres, has a radius of $500\,\text{km}$, $h_{\max}/R \sim 1$, and its form is far from being a sphere.

## Literature

H. Karttunen et al., *Fundamental Astronomy* (Springer, Berlin, 1996)

C.E. Rolfs, W.S. Rodney, *Cauldrons in the Cosmos* (University of Chicago Press, Chicago, 1988)

Victor Frederick Weisskopf. Modern physics from an elementary point of view. Lectures in the CERN Summer Vacation programme, 1969. Geneva: CERN (1970)

# Chapter 16
# Elementary Particles – Fundamental Interactions

*Science is always wrong:*
*it never solves a problem*
*without creating ten more.*

George Bernard Shaw

The great success of physics, giving us the illusion that we are indeed able to discover the secrets of nature, is probably based on our ability to explain the properties of complex systems in terms of the interactions between a few basic building blocks. This reductionist path has led us to our contemporary understanding of elementary particles and their interactions, which is described in an elegant way in terms of the standard model of elementary particles.

The more that the standard model describes the totality of particle phenomena, the more pressingly new questions pose themselves. Unless we can answer them, we will not believe that we really understand particle physics. At present the mechanism responsible for the generation of the masses of elementary particles has been experimentally confirmed. This is an important result demonstrating that the underlying theory of the standard model is renormalisable. It does not tell anything where the masses of the particle came from. How right George Bernard Shaw was with his ironic remark about science.

## 16.1 Families of Particles

The coupling of W bosons through the weak interaction to both leptons and quarks leads to us organising elementary particles into families. The decays of free W bosons have been investigated in detail in $\bar{p}p$ and in $e^+e^-$ colliders.

© Springer-Verlag GmbH Germany 2017
B. Povh and M. Rosina, *Scattering and Structures*,
Graduate Texts in Physics, DOI 10.1007/978-3-662-54515-7_16

## 16.1.1   W± *Boson Decays*

The electron–positron collider, LEP, at CERN was operated for several years at a centre of mass energy of $\approx$180 GeV. In 2000, the energy was even increased to $\approx$200 GeV. These conditions made it possible to create large numbers of pairs of W bosons, the mass of which is $M_W = (80.22 \pm 0.26)\,\text{GeV}/c^2$.

Let us first consider the decays of weak bosons into lepton pairs,

$$W^- \rightarrow e^- \overline{\nu}_e, \ \mu^- \overline{\nu}_\mu, \ \tau^- \overline{\nu}_\tau , \tag{16.1}$$

and W$^+$ into positively charged leptons and their neutrinos. The decays of W$^\pm$ display an important characteristic of the weak interaction: they arrange the leptons into three families, each consisting of a charged lepton and the corresponding neutrino $(e\nu_e)$, $(\mu\nu_\mu)$, $(\tau\nu_\tau)$. As far as we can tell from experiments, W$^\pm$ always decay into lepton pairs of the same family (Fig. 16.1). For leptons, it is usual to sort the neutrinos into families according to their flavours, as we have done here, in contradistinction to sorting them according to their mass eigenstates (Sect. 16.1.4).

On the contrary, we arrange quarks according to their mass eigenstates. Therefore, as well as the dominant decay into quarks of the same family, there are also decays into quark–antiquark pairs, in which the pairs are from neighbouring or even remote families. These decays are suppressed compared with the dominant one (Fig. 16.2).

N. Cabibbo noticed in 1963 that hadronic weak decay amplitudes are unitarily related. If one takes into account that, in the majority of experiments, the decay of

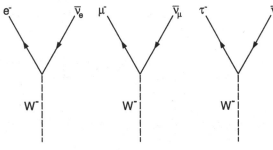

**Fig. 16.1** Decays of a W$^-$ boson into lepton pairs

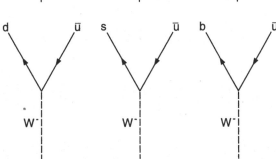

**Fig. 16.2** Decays of a W$^-$ boson into quark pairs. The decays are into quark pairs from the same (*left*), neighbouring (*middle*) and remote (*right*) families

leptons into their own neutrino partner of the same flavour is observed, then it is convenient to express this unitary relation as

$$M(\mu \to \nu_\mu) : M(n \to p) : M(\Lambda \to p) = 1 : \cos \theta_C : \sin \theta_C . \qquad (16.2)$$

Glashow, Iliopoulos and Maiani later postulated the, then unknown, charmed quark as a partner of the strange quark, to complete unitarity in two-quark families. Kobayashi and Maskawa postulated a third family of particles and increased the unitarity matrix to a 3 × 3 matrix. They did this to incorporate the, then already known, CP violation into the mixing matrix.

The unitary transformation in terms of the so-called Cabibbo–Kobayashi–Maskawa or CKM matrix relates the quark eigenstates of the mass operator (d, s, b) to a new set of quark states (d′, s′, b′), the eigenstates of the weak interaction. "Mass operator" is just a fancy phrase for the mass term in the Dirac equation,

$$\begin{pmatrix} d' \\ s' \\ b' \end{pmatrix} \approx \begin{pmatrix} 1 - \frac{\lambda^2}{2} & \lambda & A\lambda^3(\rho - i\eta) \\ -\lambda & 1 - \frac{\lambda^2}{2} & A\lambda^2 \\ A\lambda^3(1 - \rho - i\eta) & -A\lambda^2 & 1 \end{pmatrix} \begin{pmatrix} d \\ s \\ b \end{pmatrix} . \qquad (16.3)$$

In the unitary matrix (16.3), $\lambda \approx \sin \theta_C \approx 0.2$, which corresponds to the following relation: $1 - \lambda^2/2 \approx \cos \theta_C \approx 0.98$. The parameter $A$ is a real number, $\approx 0.8$. The phase $(\rho - i\eta)$ takes the small $CP$ violation in $K^0$–$\overline{K}^0$ and $B^0$–$\overline{B}^0$ systems into account, as we will see below. We have chosen this approximation for the matrix (16.3) in terms of the parameter $\lambda$ in order to bring out just how weak the mixing of hadrons (eigenstates of the mass operator versus eigenstates of the weak interaction) is! By convention, the d, b and s quarks are viewed as a superposition of d′, s′ and b′. It could be done in the same way with the u, c and t quarks.

The CKM matrix may be interpreted as follows: W bosons, strictly speaking, only couple to the weak charges. However, the quark eigenstates of the mass operator are not eigenstates of the weak interaction! These are the (u, d′), (c, s′) and (t, b′) pairs. Clearly, one cannot simultaneously diagonalise the quarks of the weak interaction and of the mass operator.

Quark decays are summarised in Fig. 16.3.

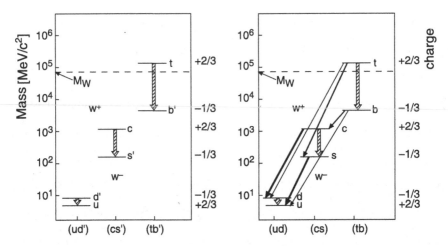

**Fig. 16.3** Quark transitions mediated by charged currents (virtual $W^\pm$ bosons): (*left*) transitions between weak interaction eigenstates; (*right*) between physical quarks. Only the t quark has a mass big enough for it to emit a real W boson. The wide arrows denote faster transitions (inside families) while the thin ones indicate the less probable transitions between families

## 16.1.2   Parity Violation and Weak Isospin

Parity violation has been investigated in detail in nuclear $\beta$ decay and in pion and muon decays. One can summarise that W bosons only couple to left-handed fermions and to right-handed antifermions. For the weak interaction, right-handed fermions and left-handed antifermions do not exist.

Each family of left-handed quarks as well as leptons forms a doublet of fermions, which can transform into each other through emission or absorption of W bosons. The electric charge of the fermions in a doublet differs by just one unit, $e$. If one only considers the weak interaction – i.e., ignores the sizeable mass differences between the fermions in a family – it is natural to view the two fermions in a doublet as two projections ($T_3 = \pm 1/2$) of a particle with weak isospin, $T = 1/2$. For right-handed antifermions, the sign of $T_3$ and the charge are both reversed (Table 16.1).

Right-handed fermions and left-handed antifermions on the other hand do not couple to W bosons and are therefore described as singlets ($T = T_3 = 0$). Left-handed leptons and the Cabibbo rotated left-handed quarks in each family thus form two doublets. If the introduction of weak isospin should have any physical meaning, then – as in the case of normal spin – $W^\pm$ are two projections of the weak isospin triplet. There has to be a third state, with $T = 1, T_3 = 0$, which should couple with the same strength, $g$, as $W^\pm$ to the fermion doublet. We denote this state by $W^0$.

**Table 16.1** Electro-weak interaction multiplets. The quarks d′, s′ and b′ are generated from the mass eigenstates by a generalised Cabibbo rotation (CKM matrix). Doublets of weak isospin $T$ are placed in brackets. The electric charge $ze$ of the two states in a doublet always differ by a unit. The sign of the third component $T_3$ is defined so that the difference $z - T_3$ inside a doublet is constant

|  | Fermion multiplets | | | $T$ | $T_3$ | $z$ |
|---|---|---|---|---|---|---|
| Leptons | $\binom{\nu_e}{e}_L$ | $\binom{\nu_\mu}{\mu}_L$ | $\binom{\nu_\tau}{\tau}_L$ | 1/2 | +1/2 −1/2 | 0 −1 |
|  | $e_R$ | $\mu_R$ | $\tau_R$ | 0 | 0 | −1 |
| Quarks | $\binom{u}{d'}_L$ | $\binom{c}{s'}_L$ | $\binom{t}{b'}_L$ | 1/2 | +1/2 −1/2 | +2/3 −1/3 |
|  | $u_R$ | $c_R$ | $t_R$ | 0 | 0 | +2/3 |
|  | $d_R$ | $s_R$ | $b_R$ | 0 | 0 | −1/3 |

## 16.1.3  $K^0$–$\overline{K}^0$, $B^0$–$\overline{B}^0$ *Oscillations and CP Violation*

$K^0$ and $\overline{K}^0$ mesons are produced via strong interactions. They both decay, through weak interactions, into two or three pions. The weak interaction couples the two mesons through the exchange of virtual pions:

$$K^0 \longleftrightarrow \begin{Bmatrix} 2\pi \\ 3\pi \end{Bmatrix} \longleftrightarrow \overline{K}^0 .$$

The probability amplitude oscillates between $K^0$ and $\overline{K}^0$.

We now want to study the time dependence in the oscillation-free basis, i.e., we have to find the eigenstates of the weak interaction. The particles and antiparticles have the same masses; however, the coupling between $K^0$ and $\overline{K}^0$ from the weak interaction, $H' = \langle K^0|\hat{H}'|\overline{K}^0\rangle = \langle \overline{K}^0 |\hat{H}'|K^0 \rangle$ breaks this mass degeneracy. The eigenstates that diagonalise $H'$ are then

$$|K_1^0\rangle = \frac{1}{\sqrt{2}}(|K^0\rangle + |\overline{K}^0\rangle)$$

$$|K_2^0\rangle = \frac{1}{\sqrt{2}}(|K^0\rangle - |\overline{K}^0\rangle) , \tag{16.4}$$

and the corresponding energies are $E_K \pm H'$. The weak interaction violates parity; indeed, it does it maximally. It also violates charge conjugation, C, which transforms particles into their antiparticles. $K_1^0$ has positive CP symmetry and decays into two pions, while $K_2^0$, with negative CP symmetry, decays into three pions. The two-pion decay mode is much quicker ($\tau \approx 10^{-9}$ s) than the three-pion decay ($\tau \approx 10^{-7}$ s), so they can both be told apart experimentally via their different decay times.

There are numerous, well known examples of oscillations in two-state systems. The reason why we are considering $K^0 - \overline{K}^0$ and $B^0 - \overline{B}^0$ here is that CP symmetry violation is only observed, albeit weakly, in these two systems. Cosmologists need a CP symmetry-violating interaction to explain the particle–antiparticle asymmetry in the universe.

The longer lived state, which can be very well measured, is, though, not a pure $K_2^0$ state. As well as three-pion decays, two-pion decays are also seen. Experiments detect kaons with an unsharp CP quantum number! These states may be written as a superposition of $K_1^0$ and $K_2^0$,

$$|K_S\rangle = \frac{1}{\sqrt{1 + \epsilon^2}}(|K_1^0\rangle + \epsilon|K_2^0\rangle)$$

$$|K_L\rangle = \frac{1}{\sqrt{1 + \epsilon^2}}(|K_2^0\rangle + \epsilon|K_1^0\rangle). \qquad (16.5)$$

This description (16.5) is only correct if the product of CPT (charge conjugation, parity and time reversal) is conserved. This is experimentally confirmed. The mixing parameter $\epsilon$ is a complex number, where $\mathrm{Re}\ \epsilon = (1.67 \pm 0.08) \cdot 10^{-3}$.

This phenomenological description of CP violation in neutral kaon systems in terms of the parameter $\epsilon$ leaves the question of the origin of this symmetry violation completely open.

Inside the standard model, there is only one source of CP violation: the complex phase of the CKM matrix. Because the mixing of the eigenstates of the weak interaction and of mass operator is described, for the antiquarks, by the complex conjugated matrix $V^*_{\mathrm{CKM}}$, the weak interaction amplitudes for antiquarks are complex conjugate to those of the quarks.

If only one amplitude $A$ contributes to the weak decay of a hadron, $h \to f$, then the observed rate for its antiparticle must be identical because $|A|^2 = |A^*|^2$. If, though, several amplitudes with different phases contribute,[1] then the particle and antiparticle rates differ due to interference terms, and phase differences can be measured.

At the end of the 1980s, the ARGUS and CLEO experiments observed strong mixing in neutral B mesons, too. In comparison with the neutral K mesons, the branching rates to the same final states are small. The mixing takes place via so-called box diagrams. The rate is proportional to the square of the mass of the top quark and is thus rather large. For CP eigenstates, $f_{\mathrm{CP}}$, which are common final states of $B^0$ and $\overline{B}^0$; interference between the amplitudes $B^0 \to f_{\mathrm{CP}}$ and $B^0 \to \overline{B}^0 \to f_{\mathrm{CP}}$ should also show a CP asymmetry between particles and antiparticles. The great advantage of this measurement is that this asymmetry offers a way to directly measure the phase differences of the CKM matrix elements.

For this reason, in the 1990s, several so-called B factories were built. Over 30 years after the discovery of CP violation, in 2000, the BaBar and Belle experiments succeeded in seeing the asymmetry in the decay $B \to J/\psi K_S$. Since then, both exper-

---

[1] The amplitudes must additionally differ by further phases, e.g., because of the strong interaction.

iments have, with ever greater statistics, explored a variety of final states. For the first time, the origin of the symmetry violation can be studied in this way. At present, all measurements agree, within errors, with the predictions of the standard model.

### 16.1.4 Neutrino Oscillations

The three neutrinos, $\nu_e, \nu_\mu, \nu_\tau$, have been experimentally detected in inverse reactions. It has also been demonstrated that they come in various flavours. All three neutrinos couple to W bosons with the universal coupling constant $g_W$. These results and the assumption that neutrinos are massless led people to conclude that $\nu_e, \nu_\mu, \nu_\tau$ are not just eigenstates of the weak interaction but also eigenstates of the mass operator. For massless neutrinos, of course, any mixture of neutrinos is also an eigenstate of the mass operator. As we shall see later, this assumption has been overthrown.

The above-mentioned experiments were, though, carried out in the immediate vicinity of the neutrino creation location in either accelerators or nuclear reactors. Measurements of solar neutrinos through inverse beta decay on $^{37}$Cl and $^{71}$Ga give a different result. On Earth, only a third to half of the flux of solar $\nu_e$ predicted by a solar models is observed. The correctness of predictions of the solar models was confirmed by measuring the neutrino flux via $Z^0$-exchange interactions, which couple to all three neutrino flavours.

The Sudbury Neutrino Observatory (Canada) detects solar neutrinos in a Čerenkov detector 2,000 m below the surface of the earth, which is filled with 1,000 tonnes of heavy water ($D_2O$). In this detector, the following reactions can be measured:

$$\nu_e + d \rightarrow p + p + e^-$$
$$\nu_{e,\mu,\tau} + d \rightarrow p + n + \nu_{e,\mu,\tau}$$
$$\nu_{e,\mu,\tau} + e^- \rightarrow e^- + \nu_{e,\mu,\tau}.$$

The first reaction only measures $\nu_e$ because the energy of the neutrino is too small to produce $\mu$ or $\tau$. The second reaction is flavour independent and measures the total neutrino flux. A total flux three times the size of the $\nu_e$ flux is indeed observed. Scattering of electrons actually has a larger cross-section for $\nu_e$ (Z and W exchange) than for $\nu_\mu$ and $\nu_\tau$ (Z exchange alone), but it does offer an additional test.

Solar neutrino oscillations imply that quantum coherence can be observed at the sun–Earth separation scale. Two properties of neutrinos follow from this: neutrino masses are nonzero and $\nu_e, \nu_\mu, \nu_\tau$ are not eigenstates of the mass operator. We denote the eigenstates of the mass operator by $\nu_1, \nu_2, \nu_3$. In analogy with quarks, we may also write the neutrino eigenstates of the weak interaction as a superposition of the neutrinos of the mass operator. For neutrinos, the matrix that is analogous to the CKM matrix should probably be called the Pontecorvo–Maki–Nakagawa–Sakata matrix. B. Pontecorvo was the first to consider the possibility of neutrino–antineutrino oscillations. The others investigated flavour mixing of neutrinos.

The unitary transformation of the Pontecorvo–Maki–Nakagawa–Sakata or PMNS matrix relates the neutrinos of the weak interaction, $\nu_e$, $\nu_\mu$, $\nu_\tau$, to a new set of neutrinos, the eigenstates of the mass operator, $\nu_1$, $\nu_2$, $\nu_3$.

The individual elements of the PMNS matrix are measured in different experiments. We will only consider two of them: oscillations in reactor antineutrinos ($\overline{\nu}_e$) in the KamLAND experiment and oscillations of $\nu_\mu$ in measurements of atmospheric neutrinos. In both cases, it suffices to consider the mixing of just two neutrino flavours.

Let us take the example of oscillation behaviour of reactor antineutrinos. Because the antineutrino energies lie far below the production threshold of muon and tau leptons, we can solely detect $\overline{\nu}_e$. This implies that we must measure the probability that antineutrinos after a displacement $L$ are in their original flavour. The time-dependent wave function of the antineutrinos is

$$|\overline{\nu}_e(t)\rangle = U_{e1}e^{-iE_{\overline{\nu}_1}t/\hbar}|\overline{\nu}_1\rangle + U_{e2}e^{-iE_{\overline{\nu}_2}t/\hbar}|\overline{\nu}_2\rangle. \tag{16.6}$$

Because the antineutrinos are relativistic, their energies can be approximated by $E_{\overline{\nu}_i} = \sqrt{p^2c^2 + m_i^2 c^4} \approx pc(1 + m_i^2 c^4/2p^2c^2)$. The probability that they still have their original flavour after a time $t$ is

$$\begin{aligned} P_{\overline{\nu}_e}(t) &= |\langle\overline{\nu}_e(t)|\overline{\nu}_e(0)\rangle|^2 \\ &= |U_{e1}|^4 + |U_{e2}|^4 + 2|U_{e1}|^2|U_{e2}|^2 \cos\left(\frac{1}{2}\frac{(m_1^2 - m_2^2)c^4}{\hbar pc}t\right). \end{aligned} \tag{16.7}$$

The oscillation length, $L_{2\pi}$, is the length at which the phase becomes $2\pi$. Letting $\Delta m_{21}^2 = m_2^2 - m_1^2$ and $t = L_{2\pi}/c$, one thus has

$$L_{2\pi} = 4\pi\frac{\hbar pc^2}{\Delta m_{21}^2 c^4} \approx 4\pi\frac{\hbar cE_{\overline{\nu}}}{\Delta m_{21}^2 c^4}. \tag{16.8}$$

In Kamioka, in Japan, reactor antineutrinos and their energies are measured in a detector of 1,000 tonnes of liquid scintillator through the reaction

$$\overline{\nu} + p \rightarrow e^+ + n, \qquad n + p \rightarrow D + \gamma + 2.2\,\text{MeV}. \tag{16.9}$$

The average separation of the detector and the reactors is $L \sim 180$ km, and the detector is sensitive to antineutrinos with energies $> 1.8\,\text{MeV}$, while the antineutrinos' energy spectrum has its peak at around $4\,\text{MeV}$. Under these conditions, as can be easily checked from (16.8), an entire oscillation length could be probed and the following oscillation parameters measured (as yet with large uncertainties): $\Delta m_{21}^2 \sim (9.0\,\text{meV}/c^2)^2$ and $U_{e1} \sim 0.84$ plus $U_{e2} \sim 0.54$.

The $\nu_\mu$ oscillations have been observed in a broad band of energies around 1 GeV. The rate at which atmospheric $\nu_\mu$'s are detected in detectors on earth, depends strongly on whether the neutrinos first pass through the atmosphere alone or through the entire earth. These observations are made at the Super-Kamiokande detector (Japan): a Čerenkov detector filled with 32,000 tonnes of water, which is 1,000 m below the surface of the earth.

Atmospheric neutrinos and antineutrinos are produced in the following decays:

$$\pi^+ \rightarrow \mu^+ + \nu_\mu$$
$$\mu^+ \rightarrow \overline{\nu}_\mu + e^+ + \nu_e,$$

and through the corresponding antiparticle decays. Initially, when the neutrinos only have to traverse the atmosphere, the ratio of muonic and electron (anti)neutrinos is $[n(\nu_\mu) + n(\overline{\nu}_\mu)]/[(\nu_e) + n(\overline{\nu}_e)] = 2$. In contrast, the flux of $\nu_\mu + \overline{\nu}_\mu$ that pass through the Earth is lower by a factor of two, although the Earth is so transparent to neutrinos that the neutrino flux should not be noticeably lessened by weak-interaction reactions. On the other hand, the flux of atmospheric $\nu_e$'s is not altered for energies of the order of GeV at the scale of the Earth's diameter.

The analysis yields $\Delta m_{32}^2 \approx (46 \, \text{meV}/c^2)^2$, with a hint that the observed $\nu_\mu$ oscillation takes place between $\nu_2$ and $\nu_3$.

In Fig. 16.4, we sketch the mass spectra of neutrinos. Because we know neither the absolute masses of the neutrinos nor the sign of $\Delta m_{32}^2$, we cannot pin down the mass scales or indeed a unique ordering of the states. From the values of $\Delta m_{21}^2$ and $\Delta m_{32}^2$ we conclude for normal ordering, that if the mass $m_1$ is very small then $m_2 \geq 9$ meV/$c^2$ and $m_3 \geq 55$ meV/$c^2$. The experimentally determined PMNS matrix can be neatly written as

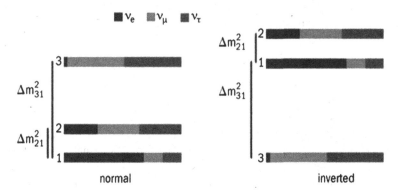

**Fig. 16.4** Because the sign of $\Delta m_{32}$ is unknown, the opposite order of the neutrino masses (*right*) cannot be excluded. The bounds on the mass of the heaviest neutrino are probably $0.05 \leq m_\nu \leq 1 \, \text{eV}/c^2$. The *shaded areas* indicate the content of flavour eigenstate in the mass eigenstates

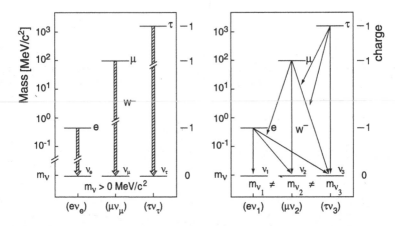

**Fig. 16.5** Charged current transitions between leptons, mediated by virtual $W^{\pm}$ bosons: *(left)* transitions between weak interaction eigenstates; *(right)* between mass operator eigenstates. The *arrows* in the *right-hand* part of the figure indicate the mixing of the mass eigenstates

$$
\begin{pmatrix} \nu_e \\ \nu_\mu \\ \nu_\tau \end{pmatrix} \approx \begin{pmatrix} \sqrt{\frac{2}{3}} & \sqrt{\frac{1}{3}} & \epsilon \\ -\sqrt{\frac{1}{6}} & \sqrt{\frac{1}{3}} & \sqrt{\frac{1}{2}} \\ \sqrt{\frac{1}{6}} & -\sqrt{\frac{1}{3}} & \sqrt{\frac{1}{2}} \end{pmatrix} \begin{pmatrix} \nu_1 \\ \nu_2 \\ \nu_3 \end{pmatrix} . \tag{16.10}
$$

The values we quote in (16.10) lie within the experimental errors. The matrix (16.10) demonstrates very clearly that the mixing between neutrinos – in contrast with hadrons – is very strong. The exception is $U_{e3} = \epsilon \approx 0.15$. Since it is nonzero, other matrix elements are slightly smaller to satisfy unitarity.

The observed lepton decays are summarised in Fig. 16.5.

## 16.2   Weak Quark Decays

Weak decays take place through virtual W boson exchange. A quark changes its flavour, i.e., its charge and perhaps its family, by emitting a virtual W boson. What happens to this depends on the phase space available to the many-particle final states. The lifetime of the quark definitely depends on the mass difference of the quarks participating in the interaction and the environment that they are in before and after the decay. This explains the wide spread of lifetimes in weak decays, which is particularly impressive in the case of nuclear $\beta$ decays. The only weak decay of which the lifetime is not dominated by the many-particle phase space, and may thus be calculated *on the back of an envelope*, is that of the top quark. Due to its large mass

$(m_t c^2 = (173 \pm 1)\,\mathrm{GeV})$ the decay into a b-quark, accompanied by the emission of a real $W^+$ ($t \to b + W^+$), is possible. This channel in fact accounts for almost 100% of the total transition probability of the decay.

## 16.2.1 Top Quark Decay

The top quark lifetime can, as usual, be estimated using Fermi's second golden rule,

$$\Gamma = \frac{2\pi}{\hbar} |\mathcal{M}|^2 \frac{4\pi p_b^2 dp_b n_s}{(2\pi\hbar)^3 dE_0} . \tag{16.11}$$

Here, $n_s$ is a spin factor that takes the three polarisation projections of the W bosons into account; $E_0 = E_b + E_W$ is the total energy of the decay. In the top system, $p_b = p_W$, and for the decay energy, one can write $dE_0 = (v_b + v_W)dp_b$. In (16.11), we replace the differential $dE_0/dp_b = v_b + v_W$ and obtain the expression for the transition probability with the final form of the phase space,

$$\begin{aligned}
\Gamma &= \frac{2\pi}{\hbar} |\mathcal{M}|^2 \frac{4\pi p_b^2 n_s}{(2\pi\hbar)^3 (v_b + v_W)} \\
&= \frac{2\pi}{\hbar} |\mathcal{M}|^2 \frac{4\pi p_b^2 n_s}{(2\pi\hbar)^3 p_b c^4 (E_b + E_W)/E_b E_W} .
\end{aligned} \tag{16.12}$$

In our simplified electro-weak unification ($\alpha_W \sim \alpha$), we can approximate the matrix element by $\mathcal{M}^2 \approx 4\pi\alpha(\hbar c)^3/(2E_W)$. The sum $E_b + E_W = m_t c^2$ is the top mass and $E_b \approx p_b c$,

$$\Gamma \approx 2\alpha \frac{p_b^2}{m_t} n_s . \tag{16.13}$$

The estimates $p_b^2 \approx \frac{1}{6}(m_t c)^2$ and $n_s \approx 3$ yield

$$\Gamma \approx \alpha m_t c^2 . \tag{16.14}$$

The elementary electro-weak decay width corresponds to a vertex in a Feynman graph and typically its value is 1/137 of the mass of the decaying particle.

An exact calculation using Glashow, Weinberg and Salam's electro-weak theory yields almost exactly the same result. Instead of $\alpha$, one has to use the weak coupling, $f_{tb}^2 \alpha_W = \frac{1}{4}\alpha/\sin^2\theta_W = 1.081\alpha$, where $f_{tb} = \frac{1}{2}$ is the matrix element for the $t \to b$ transition. The spin factor is a little larger than 3, $n_s = \frac{1}{2} + \frac{1}{2} + \frac{1}{2}(m_t/m_W)^2 = 3.34$; due to the averaging over $\sin^2(\theta/2)$, both the transverse components only contribute a factor of one half, while the longitudinal component dominates. The phase-space factor is $p_b^2/(m_t c)^2 = [1 - (m_W/m_t)^2]^2 = 0.155$ instead of 1/6. The weak and strong radiative corrections introduce a factor of 1.02. These factors lead to

$$\Gamma = 1.14\alpha m_t c^2 = 1.45\,\text{GeV}, \tag{16.15}$$

which favourably compares to the experimental value 1.41 GeV. This decay width corresponds to a lifetime $\tau = \hbar/\Gamma = 0.5 \cdot 10^{-24}$s, which may seem rather short, but compared with the time scale for the top quark, $\hbar/(m_t c^2)$, it is very long, $\tau = 137\hbar/(m_t c^2)$.

## 16.3   $Z^0$ and the Photon

The $Z^0$ boson is not the $W^0$ boson, which we predicted from our considerations of weak isospin. The mass of the $Z^0$ boson is $(91.188 \pm 0.007)\,\text{GeV/c}^2$, which is almost $11\,\text{GeV/c}^2$ larger than the masses of the W bosons. Because we still understand little about the masses of the particles, this difference in the masses is not a strong argument against our claim. But the decay of the $Z^0$ boson unambiguously shows that it does not only couple weakly to fermions (Fig. 16.6).

Analysis of experimental data from LEP and SLAC (see the Particle Data Group) yields the following branching ratios:

$$
\begin{aligned}
Z^0 \longrightarrow \quad & e^+ + e^- & 3.363 \pm 0.008\% \\
& \mu^+ + \mu^- & 3.366 \pm 0.013\% \\
& \tau^+ + \tau^- & 3.370 \pm 0.015\% \\
& \nu_{e,\mu,\tau} + \bar{\nu}_{e,\mu,\tau} & 20.00 \;\pm 0.16\;\% \\
& \text{hadrons} & 69.91 \;\pm 0.15\;\% .
\end{aligned}
\tag{16.16}
$$

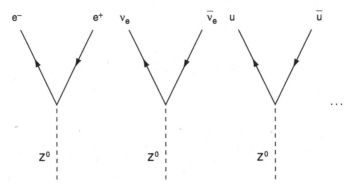

**Fig. 16.6** Decay of the $Z^0$ boson into fermion pairs. Fermions and antifermions always have the same flavour

It is clear that $Z^0$ decay distinguishes between charged leptons and neutrinos as well as between quarks with different charges. If all fermion pairs coupled in the same way to $Z^0$, then one would expect $1/21$ for each lepton pair and $15/21$ for hadrons (due to the three colour charges and five active quark flavours). We are already accustomed to this: the mass generating interaction casts the elegant symmetries of the weak interactions into disarray. The original gauge bosons of the weak and electromagnetic interactions are the three $W^{\pm,0}$ bosons of the weak isospin SU(2) symmetry and the original photon B of U(1) symmetry.

The perfect SU(2)×U(1) symmetry is broken by the mass generating inter-action, which causes states to mix in a fashion similar to CKM mixing. The experimental photon and $Z^0$ are related to the original photon B and $W^0$ via a unitary transformation.

The unitary transformation between the initial photon B and $W^0$,

$$\begin{aligned}
|\gamma\rangle &= \quad \cos\theta_W |B\rangle + \sin\theta_W |W^0\rangle \\
|Z^0\rangle &= -\sin\theta_W |B\rangle + \cos\theta_W |W^0\rangle,
\end{aligned}$$

is expressed via the so-called Weinberg angle, $\theta_W$.

This mixing also causes a mixing of the weak decay amplitude (the third compo-nent, $T_3$, of weak isospin) with the electromagnetic amplitude (the electric charge $z$), so that the partial decay widths are proportional to $(T_3 - z\sin^2\theta_W)^2$. Right-handed fermions do not have a weak coupling, $T_3 = 0$. Left-handed negative leptons and quarks have weak isospin, $T_3 = -1/2$, while neutrinos and positive quarks have $T_3 = +1/2$. Adding up the left-handed and right-handed contributions and approx-imating $\sin^2\theta_W \approx 1/4$, we have

$$\Gamma \propto (-z\sin^2\theta_W)^2 + \left(\frac{1}{2} + |z|\sin^2\theta_W\right)^2 \approx \frac{1}{8}\left(2 - 2|z| + z^2\right), \qquad (16.17)$$

and, for $Z^0$ decay, we have the approximate ratios

$$\Gamma(e^+e^-):\Gamma(\nu_e\bar{\nu}_e):\Gamma(u\bar{u}):\Gamma(d\bar{d}) \approx 1:2:\left(3\times\frac{10}{9}\right):\left(3\times\frac{13}{9}\right), \qquad (16.18)$$

and similar for the second and third families. The factor of 3 is a result of the three quark colours (see Chap. 3). The ratios for the more exact value $\sin^2\theta_W = 0.2312$ are $1 : 1.99 : 3.42 : 4.41$. Experimentally, they are $1 : 1.98 : 3.00 : 4.93$, where we have taken the average of the three families (for $u\bar{u}$ and $c\bar{c}$ pairs; we actually take only two families because the $t\bar{t}$ pair is too heavy to be produced). The agreement for leptons is excellent, while the 10% level disagreements for quarks indicates the effects of other physical influences and phenomena. The final state does not, due to confinement, consist of free quarks because they hadronise and the available phase-space influences the decay probabilities.

## 16.4  Higgs Ex Machina

The idea of the Higgs field was introduced in order to rescue the Standard Model. In the Standard Model infinite integrals appear and a regularization and renormalisation is needed. It has been proven that for the renormalisability the Lagrangian has to be gauge invariant which, at the face value, requires massless fermions and massless gauge bosons. An explicit bilinear mass term in the Lagrangian would spoil gauge invariance. The problem can be solved if instead of the mass term one introduces a coupling of the gauge bosons and fermions to the scalar field – Higgs field, whose one component has non-zero vacuum expectation value.

In the following three figures we sketch the Higgs mechanism. In Fig. 16.7 the massless gauge boson of the electroweak interaction $W_1$, $W_2$, $W_3$ and $B$ are shown.

In order to couple correctly the Higgs field to fermions and weak bosons it must consist of two SU(2) doublets which are then rearranged according to Fig. 16.8.

The two charged components of the Higgs field get absorbed in the longitudinal components of $W_1$ and $W_2$ bosons as indicated in Fig. 16.9. The neutral component $H^0$ mixes the $W_3$ and $B$ fields and it contributes to the longitudinal component of $Z^0$ bosons and it leads to the massless photon. The fourth component acts as a mass term. Its fluctuation around the vacuum value represents an observable particle. Luckily, the Higgs boson has been experimentally confirmed, at a mass of 125 GeV/c$^2$.

According to George Bernard Shaw, our solution of one problem has created 10 new ones (although we are very happy about it). We here only mention three, which, though, are no simpler than the problem we have solved:

(i) Which mechanism is responsible for the actual values of the masses of the elementary particles? The Higgs mechanism by itself does not answer this question. The particle masses are taken out from the experiments and are in the standard model just free parameters. For their origin even a theoretical concept is still lacking.

**Fig. 16.7** The massless gauge bosons of the electroweak interaction

**Fig. 16.8** The four states of the Higgs boson. The first three give mass to the weak bosons and leave the photon massless. The fourth state is an observable particle

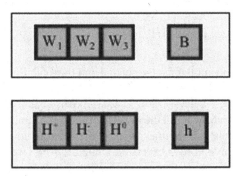

**Fig. 16.9** Cartoon
visualising the Higgs model

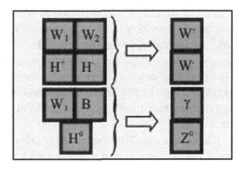

(ii) What produces the conflict between the mass operator and the weak interaction so that the mass eigenstates, d, s, b, are not the same as the weak doublet partners, d′, s′, b′? We do not know, but we also do not know any symmetry that could stop it. In nature, everything is allowed – we think – that is not explicitly prohibited by a conservation law.

(iii) How heavy are neutrinos? In the simplified standard model, one assumes that neutrinos are massless. There is, though, experimental evidence of a CKM-style mixing for solar and atmospheric neutrinos indicating small but finite neutrino masses of the order of a few meV.. The question is what mechanism is capable to produce such small masses.

In a naive electro-weak theory with full $U(1) \times SU(2)$ symmetry, the electron and neutrino would have the same mass and electric charge. Similarly, the partners of the $SU(2)$ doublets $(\mu, \nu_\mu)$, $(\tau, \nu_\tau)$, (d, u), (s, c) and (b,t) would have degenerate masses. Because we believe that the world is described by renormalisable field theories, the weak bosons would have to be massless too. This is where the deus ex machina – the Higgs field – emerges to give the particles the correct effective masses. This takes place through a phase transition embodied by a spontaneous symmetry breaking.

Let us start to sketch the phase transition scenario. The Higgs field must have weak isospin to couple to the weak bosons. This means that it must comprise $SU(2)$ doublets. There must furthermore be at least two doublets because three components of the Higgs field must be converted into longitudinal components of the $W^\pm$ and $Z^0$ bosons; this is because massless weak bosons only have two (transverse) degrees of freedom. When they pick up a mass, though, they require an extra longitudinal degree of freedom. The Higgs field produces then the masses, the longitudinal spin components and additionally – as we will later see – the mixing between the original photon B and the $W^0$. The fourth component of the Higgs field is left over as a physical particle.

The simplest model of a phase transition may be constructed by giving the order parameter, here the Higgs field, a nonzero vacuum expectation value. This may be achieved via a Higgs field potential $V(\Phi)$ of the form shown in Fig. 16.10, in which the field $\Phi$ can take on an arbitrary value inside the ring-like minimum. This is called spontaneous symmetry breaking because the vacuum is no longer symmetric under $SU(2)$.

Such a pattern of phase transition was already applied in Chap. 6 to describe spontaneous breaking of chiral symmetry. There, too, the curve of energy against the order parameter (the constituent mass $M$) resembles a Mexican hat with a degenerate ground state (vacuum) at a nonzero value of $M$. In Chaps. 5 and 6, though, we described the phase transitions as a consequence of a feedback of the order parameter; see (5.27) and (12.17). Both descriptions are equivalent because the back-coupling equation is a variational equation for the energy surface with the form of a Mexican hat.

It is usual to describe the two Higgs doublets as a complex field, $\Phi$, such that the upper complex components correspond to a positive and a negative particle while the lower complex components represent two neutral particles. Because the vacuum is neutral, only the lower (neutral) components of the Higgs doublet can have a nonzero vacuum expectation value, which we denote by $v$. The phase of the Higgs field can always be defined such that $v$ is real. We expand the Higgs field, $\Phi$, around its vacuum expectation value, $\Phi_0$. The individual components of the four real fields correspond to fluctuations around the vacuum value.

$$\Phi_0 = \frac{1}{\sqrt{2}} \begin{pmatrix} 0 \\ v \end{pmatrix} \qquad \Phi = \frac{1}{\sqrt{2}} \begin{pmatrix} \chi_1 + i\chi_2 \\ (v + \chi_3) + i\chi_4 \end{pmatrix}. \tag{16.19}$$

The components $\chi_{1,2}$ are transformed into longitudinal components of the weak bosons $W^\pm$; $\chi_4$ becomes the longitudinal component of the neutral weak boson, $\chi_3$ corresponds to a physical particle. Why $\chi_3$? This is because $\chi_3$ accompanies $v$ and describes fluctuations of the Higgs field in the steep direction of the Mexican hat (Fig. 16.10).

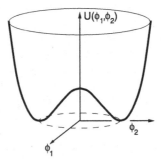

**Fig. 16.10** At low temperatures ("low-energy" phenomena below the TeV threshold), the potential's minimum is at a nonzero value of the Higgs field, which leads to spontaneous breaking of SU(2)×U(1) symmetry. At high temperatures ($kT > 2v$) and densities on the other hand, the potential looks very different: it has a minimum for $\Phi_{\text{Higgs}} = 0$, and the symmetry is restored. The co-ordinates $\phi_1$ and $\phi_2$ are the real parts of the upper and lower components of the Higgs field $\Phi$

Minimal coupling of the Higgs field to B and W bosons has the form

$$\mathcal{L}_{\text{Higgs}} = \left[ \left( ig\frac{\boldsymbol{\tau}}{2}\mathbf{W}^\mu + ig'\frac{1}{2}B^\mu \right) \frac{1}{\sqrt{2}} \begin{pmatrix} 0 \\ v \end{pmatrix} \right]^\dagger$$
$$\times \left[ \left( ig\frac{\boldsymbol{\tau}}{2}\mathbf{W}_\mu + ig'\frac{1}{2}B_\mu \right) \frac{1}{\sqrt{2}} \begin{pmatrix} 0 \\ v \end{pmatrix} \right]. \tag{16.20}$$

The $(2 \times 2)$ Pauli matrix, $\boldsymbol{\tau}$, is the weak charge and the factor $\frac{1}{2}$ in the coupling to the original photon field $B$ is the U(1) hypercharge of the Higgs field. The coupling constants $g$ and $g'$ are related to $\alpha$ and $\alpha_W$ in such a way that $\alpha_W = g^2/4\pi$, $\tan\theta_W = g'/g$ and $\alpha = \alpha_W \sin^2\theta_W$ hold.

Here, we have only written the vacuum term of the Higgs field because it is this alone that produces the quadratic terms in the $W$, $B$ and $\chi$ fields, which are important for mass creation. The complete Higgs field also contributes to cubic and quartic terms, which are responsible for various processes such as Higgs production and decay. Between the most important decays of the Higgs boson are decays into pairs of $Z^0 Z^0$ or $W^+ W^-$, which involve the full coupling $g$. Such decays may be straightforwardly detected through the decay products of the weak bosons – two pairs of jets or leptons. It is a lucky situation that lepton pairs (especially muon pairs) are easily identified. Since the Higgs mass is less than twice the mass of the weak bosons, at least one of the weak bosons must be virtual. Anyway, we are used to virtual weak bosons in beta decay.

The quadratic terms of the weak fields look like mass terms and can be interpreted as such. The bosons pick up a mass because they stick to the Higgs field. Through diagonalisation, one can get rid of the mixed $-2gg' W^{0\mu}B_\mu/4$ terms,

$$\frac{1}{4} \left( W^{0\mu},\ B^\mu \right) \begin{pmatrix} g^2 v^2, & -gg' v^2 \\ -gg' v^2, & g'^2 v^2 \end{pmatrix} \begin{pmatrix} W^0_\mu \\ B_\mu \end{pmatrix}$$
$$\rightarrow \frac{1}{4} \left( Z^{0\mu},\ A^\mu \right) \begin{pmatrix} (g^2 + g'^2)v^2, & 0 \\ 0, & 0 \end{pmatrix} \begin{pmatrix} Z^0_\mu \\ A_\mu \end{pmatrix},$$

and so obtain seen.

$$\mathcal{L}_{\text{Higgs}} = 2\frac{m_W^2 c^4}{2} W^{+\mu}W^-_\mu + \frac{m_Z^2 c^4}{2} Z^{0\mu}Z^0_\mu + \frac{m_\gamma^2 c^4}{2} A^\mu A_\mu \tag{16.21}$$

where $m_W c^2 = gv/2$, $m_Z c^2 = \sqrt{g^2 + g'^2}\, v/2 = m_W c^2/\cos\theta_W$ and $m_\gamma = 0$. We have denoted the photon field by $A$. This diagonalisation corresponds to Weinberg mixing (16.17). From $g = (e/\sqrt{\varepsilon_0 \hbar c})/\sin\theta_W = 0.6$, one can calculate the vacuum expectation value $v = 2m_W c^2/g = 246\,\text{GeV}$, although it has no measurable physical meaning.

Fermions also couple to the Higgs field and so pick up a mass. To investigate this, it suffices to use the simplest form of the coupling, the so-called Yukawa coupling (contact coupling),

$$\mathcal{L}'_{\text{Higgs}} = -\sum_\alpha \frac{g_\alpha}{\sqrt{2}} (v + \chi_3) \overline{\psi}_\alpha \psi_\alpha$$
$$= -\sum_\alpha m_\alpha c^2 \left(1 + \frac{\chi_3}{v}\right) \overline{\psi}_\alpha \psi_\alpha. \tag{16.22}$$

We have interpreted $g_\alpha v / \sqrt{2} = m_\alpha c^2$ as fermion masses. The price of such mass generation is the coupling of fermions to the Higgs field $\chi_3$. The other components $\chi_{1,2,4}$ are not described here because they may be better rewritten as coupling to the equivalent longitudinal components of the weak bosons and kept together in the Lagrange density with the transverse components in the quark-W or quark-Z coupling.

It is noteworthy that the coupling $g_\alpha / \sqrt{2} = m_\alpha c^2 / v = g(m_\alpha/m_W)$ is proportional to the fermion mass, $m_\alpha$. Therefore the dominant decay of the Higgs particle is into heavy quark–antiquark pairs, actually into $b\overline{b}$. The $t\overline{t}$ pair is too heavy and appears only in some less probable virtual processes leading between others to a noticeable $\gamma\gamma$ decay.

The Higgs model has, despite its elegance, an imperfection: each individual fermion requires a priori an arbitrary coupling constant $g_\alpha$. This is (up to the factor $\sqrt{2} c^2 / v$) equal to the mass generated by the Higgs mechanism. Where the mass of the particles comes from belongs to the *Physics beyond the standard model*.

It should be stressed that the Higgs mechanism is not the whole story. It provides only the bare mass which in most cases is not very far from the dressed mass (the constituent mass). The difference is, however, dramatic for u and d quarks which have a bare mass of only 3–7 MeV/c$^2$. (provided by the coupling to the Higgs boson) and a dressed (constituent) mass of about 330 MeV/c$^2$. The dressing is provided by the strong interaction – the gluon condensate. Pictorially, our human body weighs predominantly because of our gluon content.

## 16.5  Proton Decay

Unification has so far proved to be an important principle in physics. Newton indeed introduced his gravitational theory through the hypothesis that the same laws hold on the eartseenh and in the heavens. Maxwell showed that the electric and magnetic interactions can be explained using a single coupling constant. Nowadays, people are trying to apply this pattern of unification to the electromagnetic, weak and strong interactions in the framework of a grand unified theory.

Initially, it may seem rather unlikely that all three interactions can be described in terms of a common coupling constant, after all at currently attainable accelerator energies they are very different: $\alpha = 1/137$, $\alpha_W = 1/32$, $\alpha_s \approx 1/5 - 1/9$.

**Fig. 16.11** A clue to a possible grand unified theory follows from extrapolating the running coupling constants $\alpha_s$, $\alpha_W$, $\alpha_B$. The inverse coupling constants are sketched against the scale, $Q^2 = \mu^2$, at which the couplings are measured

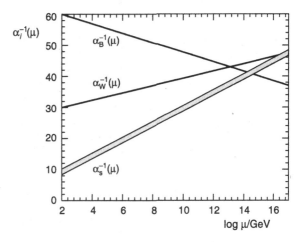

A hint that unification may, however, be possible is given by the running of the coupling constants ((3.19) and (3.20)). The the bare mass. electromagnetic coupling constant at higher resolutions (with increasing $Q^2$) is seen to get stronger: the vacuum polarisation screens charges but at smaller separations more of the charge is seen. For the weak and strong interactions this is reversed: weak bosons carry weak isospin and gluons colour, self-coupling effects outweigh those of the vacuum polarisation and both interactions become weaker at larger $Q^2$. Because this extrapolation is carried out above the weak energy scale (100 GeV) the original photon decouples from $W^0$, and the coupling constant of the original photon must be used instead of the electromagnetic one: $\alpha_B = (5/3)g'^2/4\pi = (5/3)\alpha/\cos^2\theta_W$. The factor 5/3 is due to the uniform normalisation of all three constants. At around $10^{15}$ GeV, all three coupling constants join at a value around 1/45, Fig. 16.11.

This extrapolation is, of course, only meaningful when there is no new physics between the weak scale, 100 GeV, and the unification scale of $10^{15}$ GeV.

The underlying idea of grand unified theory is that there is a phase transition at this energy scale to a larger symmetry and that transitions from quarks into leptons are possible. The exchange bosons associated with these transitions are called X bosons. Their masses are roughly the unification scale.

An experimental test of the unification hypothesis is provided by proton decay. An *on the back of an envelope* estimate of the proton lifetime may be carried out if we assume that the X bosons have a mass $m_X = 10^{15}$ GeV/$c^2$. The simplest thing is to compare with a weak decay, which has the same phase space as is expected for proton decay

Let us compare a decay channel of the proton (p $\rightarrow \pi^0 + e^+$) with that of a weak decay of the D meson ($D^+ \rightarrow \overline{K}^0 + \pi^+$) (Fig. 16.12). Let us further assume that the other decay channels all have roughly the same phase space. In a grand unified theory, all the coupling constants are the same, the only dramatic differences are in

**Fig. 16.12**  The phase space for proton decay is comparable with that of the Cabibbo-allowed decay of a D meson

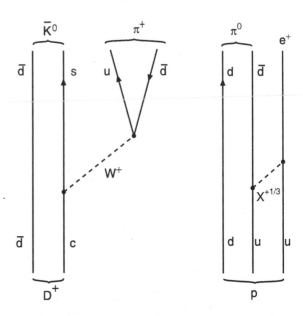

the propagators. The matrix elements are proportional to the inverse square of the boson mass:

$$\mathcal{M}(\mathrm{p} \to \pi^0 \mathrm{e}^+) \propto \frac{1}{m_X^2}, \qquad \mathcal{M}(\mathrm{D}^+ \to \overline{\mathrm{K}}^0 \pi^+) \propto \frac{1}{m_W^2}, \tag{16.23}$$

which implies the lifetimes ratio

$$\frac{\tau(\mathrm{p} \to \pi^0 \mathrm{e}^+)}{\tau(\mathrm{D}^+ \to \overline{\mathrm{K}}^0 \pi^+)} \approx \left(\frac{m_X}{m_W}\right)^4 \approx 10^{52}. \tag{16.24}$$

The $\mathrm{D}^+$ lifetime is $10^{-12}$s, from which we can read off our estimate of the proton lifetime

$$\tau_{\mathrm{proton}} \approx 10^{52} \times 10^{-12}\mathrm{s} \approx 10^{40}\mathrm{s} \approx 10^{32}\mathrm{years}. \tag{16.25}$$

Up to now, experiments have not yielded firm evidence for proton decay and this has given us a lower bound on the proton lifetime, $\tau_{\mathrm{proton}} > 10^{32}$ years. Perhaps the unification scale lies at a higher energy than the extrapolation of Fig. 16.11 suggests or the present GUT model is not what nature has realised.

# Literature

Q.R. Ahmad et al., Measurement of the Rate of $v_e + d \rightarrow p + p + e^-$. Interactions produced by 8B solar neutrinos at the Sudbury neutrino observatory. Phys. Rev. Lett. **87**, 071301 (2001)

S. Eidelman et al., Review of particle physics. Phys. Lett. B **592**(1), 1–5 (2004)

H. Frauenfelder, E. Henley, *Subatomic Physics* (Prentice-Hall, Englewood Cliffs, NJ, 1991)

S. Fukuda et al., Solar 8B and hep neutrino measurements from 1258 days of super-kamiokande data. Phys. Rev. Lett. **86**, 5651–5655 (2001)

F. Halzen, A.D. Martin, *Quarks and Leptons* (Wiley, New York, 1984)

D. Perkins, *Introduction to High Energy Physcis* (University Press, Oxford, 2000)

B. Povh et al., *Particles and Nuclei* (Springer, Berlin, 2015)

# Chapter 17
# Cosmology – The Early Universe

*Evolutionary theory of gravitation: In the beginning the world
was symmetric; stones were flying in all directions; only those
falling down remained.*

anonymous

*Cosmologists are seldom right but never in doubt.*

L.D. Landau

The examples of symmetry breaking discussed in Chap. 16 may be elegantly incorporated into the standard big bang model (Fig. 17.1). In this model, the universe is, for the first few fractions of a second (time is a parameter of the model!), a genuine, if exotic, quantum system where all interactions are unified. During the first stage of the cooling of the universe, gravity separates out from the other interactions. Subsequently, the electro-weak interaction separates from the strong interaction; simultaneously, the leptons separate from the quarks. This takes place at a temperature around $10^{15}$ GeV. Some of the bosonic states ($\gamma$, $W^{\pm,0}$, gluons) remain massless, while others acquire a large mass, likely at the same scale. Up to this stage, fermions can freely transform into one another, afterward, the large masses of the exchange bosons prevent this and the decay of the proton into a positron and $\pi^0$ takes at least $10^{32}$ years.

At a temperature of around 300–100 GeV, a further symmetry breaking takes place in which the electromagnetic interaction separates from the weak one. The weak bosons acquire a mass, which corresponds to the symmetry breaking energy, and the fermions obtain additional properties. The particles inside doublets (e, $\nu_e$), (u, d), etc. thereafter differ in charge, mass, flavour and family.

Somewhere in between, there is probably a further symmetry breaking, which marks the difference between left- and right-handed fermions and establishes a preferred weak coupling to left-handed fermions.

© Springer-Verlag GmbH Germany 2017
B. Povh and M. Rosina, *Scattering and Structures*,
Graduate Texts in Physics, DOI 10.1007/978-3-662-54515-7_17

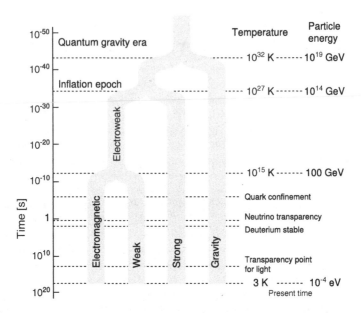

**Fig. 17.1** During the first cooling phase gravity separates out from the other interactions. Subsequently, the electro-weak interaction separates from the strong interaction and the electromagnetic interaction separates from the weak one

At a temperature around 200 MeV, chirality is broken and confinement starts up; protons and neutrons are then built out of quarks.

We see that the history of the universe after the big bang is characterised by a succession of symmetry breakings from larger to smaller energies. Finally, breaking of translational invariance in the universe happened and matter became inhomogeneous and the process of its clumping together in galaxies started.

The final significant symmetry breaking of which we are aware, or – as is normally said of this sort of thing – self-organisation in the universe, is the appearance of life and humanity on earth.

In the following, we want to mention a few of the hints from cosmology that point to the need to extend the standard model of elementary particles. These clues can, though, only come from astronomical observations and their interpretation.

## 17.1  The Three Pillars of the Big Bang Model

### 17.1.1  The Expanding Universe

Already in 1929, Hubble had observed that distant galaxies are moving away from us at speeds that are proportional to their distance, $d$, $v = H \cdot d$. The Hubble parameter,

$H$, today $(t_0)$ has a value $H(t_0) = 72 \pm 3\,\mathrm{km\,s^{-1}\,Mpc^{-1}} \approx (14 \cdot 10^9\,\mathrm{yrs})^{-1}$. The Hubble parameter, $H$, is time dependent. It is therefore sensible to use the following vector relation between position and velocity vectors with respect to an arbitrary origin:

$$\mathbf{v}(t) = H(t)\mathbf{x}(t). \tag{17.1}$$

Because, in astronomy, distances rather than times are measured, it is useful to take the time dependence of the co-ordinates into account via a scalar factor with dimension of length, which calibrates the measured distance as a function of the cosmic time scale. We thus define a scalar parameter $R(t)$ via

$$\mathbf{x}(t) = R(t)\mathbf{x}_0, \tag{17.2}$$

with $R(t_0) = R_0$, so that $\mathbf{x}(t_0) = R_0\mathbf{x}_0$ is the contemporary co-ordinate of the observer. Thus, we can put the Hubble law ansatz (17.1) into scalar form,

$$\mathbf{v}(t) = \dot{\mathbf{x}}(t) = \dot{R}(t)\mathbf{x}_0 = H(t)R(t)\mathbf{x}_0. \tag{17.3}$$

The Hubble parameter is thus the rate of change of the scalar parameter

$$H(t) = \frac{\dot{R}(t)}{R(t)}. \tag{17.4}$$

A positive value of $H$ corresponds to an expanding universe.

The Hubble parameter has dimension of inverse time, $H \approx 2.3 \cdot 10^{-18}\,\mathrm{s^{-1}}$, and the Hubble time $t_H$ can thus be defined as

$$t_H = \frac{1}{H(t_0)} \approx 14 \cdot 10^9\,\mathrm{yrs}. \tag{17.5}$$

The Hubble time yields the correct order of magnitude of the age of the universe.

The time dependence of the scaling parameter $R$ can be obtained by a simulation of the expanding universe based on the so-called Friedmann model. Friedmann was the first to appreciate, in 1922, that Einstein's equations possess cosmological solutions that only contain matter. In 1927, Lemaître found the solutions to the Friedmann equation and showed that they led to a linear distance-redshift relation. The Friedmann equation is

$$\dot{R}^2 - \frac{8\pi G}{3}\rho R^2 = -kc^2. \tag{17.6}$$

The left-hand side of (17.6) can be interpreted in the Newtonian frame as energy conservation: $(\dot{R}r)^2 - GM/(Rr) = \mathrm{constant}$. General relativity says that the constant is the sum of all energies: the energy density, $\rho$, receives contributions from matter, radiation and the vacuum. The parameter $k$ determines the curvature of space. All astronomical observations up to now are consistent with a flat universe, $k = 0$, and in the following, we will only consider this case.

The parameter $k$ does not just define the geometry of the universe but, via (17.6), it also determines the value of the average density of the universe, the so-called critical density, $\rho_c$. From (17.6), the value of the critical density is deduced to be

$$\rho_c = \frac{3}{8\pi G} \left(\frac{\dot{R}}{R}\right)^2 = \frac{3}{8\pi G} H^2(t). \tag{17.7}$$

Its numerical value expressed in terms of proton masses per cubic meter is $\rho_c \approx 5.6\, m_p/\mathrm{m}^3$.

One distinguishes between two periods in the history of the universe. The first is a radiation-dominated era, while the second is a matter-dominated era. In our present matter-dominated universe, the energy density is inversely proportional to the volume of the universe, $\rho \sim 1/R^3$. In the earlier radiation-dominated universe, the wavelength of the radiation scaled with $R$, so that taking the volume of the universe into account, $\rho \sim 1/R^4$. Equation (17.6) may be solved by substituting these two possibilities for $\rho$ into it. The scaling parameter, $R(t)$, in the matter-dominated universe is proportional to $t^{3/2}$, while, for the radiation-dominated one, it scales as $t^{1/2}$. The Hubble parameter, $H = \dot{R}/R$, is then $2/3t$ for the matter-dominated and $1/2t$ for the radiation-dominated universe. The main results of the Friedmann model are summarised in the following:

| Radiation dominated | Matter dominated |
|---|---|
| $R = R_0 \cdot \left(\dfrac{32G\rho_0}{3}\right)^{1/4} \cdot \sqrt{t}$ | $R = R_0 \cdot (6\pi G\rho_0)^{1/3} \cdot t^{2/3}$ |
| $H = \dfrac{\dot{R}}{R} = \dfrac{1}{2t}$ | $H = \dfrac{\dot{R}}{R} = \dfrac{2}{3t}$ |
| $T \propto t^{-1/2}$ | $T \propto t^{-2/3}$ |
| $\rho = \dfrac{3}{32\pi G} \cdot t^{-2}$ | $\rho = \dfrac{1}{6\pi G} \cdot t^{-2}$ |

$$\tag{17.8}$$

Because of the singularity at $t = 0$, $R$ has to be normalised at $t_0$. $H$ and $\rho$ follow from Friedman equations and the temperature $T$ follows from thermodynamics.

Observations of the expanding universe alone do not provide compelling evidence for the big bang model. It is possible, though, to look further back into history and see that the universe looked very different and less differentiated.

The further galaxies are away from us, the more quickly they are receding from us and light from them is shifted further into the red. All distances in the universe scale with the scaling factor $R(t)$, and this also applies to the wavelength of light. This leads to the relation between the frequency of emitted and observed light

$$\frac{\omega_{emit}}{\omega_{obs}} = \frac{R(t_{obs})}{R(t_{emit})} = 1 + z, \tag{17.9}$$

where $z$ is the Doppler shift. Modern observations of the furthest seen galaxies stretch to $z \approx 6$ when the universe was 5 billion years younger than today. After this, the next well-defined observation is at $z \approx 1000$, which is at a time when radiation had decoupled from matter.

### 17.1.2 Cosmic Microwave Background Radiation

The cosmic microwave background (CMB) radiation is turning out to be an extremely rich source of information about our universe at the time of the decoupling of electromagnetic radiation from matter. This relic from the early universe gives an important calibration point of the cosmic time scale. It is used to support our present belief that the universe is flat, i.e., $k$ in (17.6) is 0 and thus has the critical average density. It offers evidence for how the first structures in the universe formed.

After the "first three minutes" (see next subsection), the universe was composed of a plasma of fully ionised hydrogen and helium and about $10^{10}$ times as many photons. The main mechanism for energy transport in this period was Compton scattering. The photon mean free path was small at the cosmic scale and the universe was opaque.

One would expect that the decoupling of radiation from matter started when the temperature became too low to keep the thermal equilibrium via the reaction

$$p + e \leftrightarrow H + \gamma. \tag{17.10}$$

It is instructive, if not fully physically justified, to use the equilibrium formulae to carry out an *on the back of an envelope* calculation. As in the case of $3\alpha$ fusion (15.31), (15.33), we need the chemical potentials before and after the reaction to be equal:

$$kT \ln \frac{n_p}{2} \left(\frac{2\pi \hbar^2}{m_p kT}\right)^{3/2} + kT \ln \frac{n_e}{2} \left(\frac{2\pi \hbar^2}{m_e kT}\right)^{3/2}$$
$$= kT \ln \frac{n_H}{4} \left(\frac{2\pi \hbar^2}{m_H kT}\right)^{3/2} - Q, \tag{17.11}$$

or equivalently,

$$\ln \frac{n_p n_e}{n_H} \left(\frac{2\pi \hbar^2 m_H}{m_p m_e kT}\right)^{3/2} = \frac{-Q}{kT}. \tag{17.12}$$

Here, $Q = 13.6\,\text{eV}$ is the liberated ionisation energy and $\frac{1}{2}$ or $\frac{1}{4}$ are spin factors. It is more common to write (17.12) its exponential form (SAHA equation)

$$\frac{n_p n_e}{n_H} = \left(\frac{m_p m_n kT}{2\pi \hbar^2 m_D}\right)^{3/2} e^{-Q/kT}. \tag{17.13}$$

For our estimate, we assumed that the cross-over took place at $n_e \approx n_H$. The proton density, $n_p$, can be obtained by extrapolating its present density, $n_p(t_0) = 0.15\,\text{m}^{-3}$, to the temperature $T_{\text{dec}}$, when the recombination of hydrogen began:

$$n_p \approx n_p(t_0) \left(\frac{T_{\text{dec}}}{2.7\,\text{K}}\right)^3. \tag{17.14}$$

The decoupling temperature, $T = T_{\text{dec}}$, is obtained by inserting (17.14) into (17.13), with the solution $kT_{\text{dec}} = 0.32\,\text{eV}$ ($T_{\text{dec}} \approx 3700\,\text{K}$).

However, the recombination of hydrogen actually started later, at somewhat lower temperatures than $kT = 0.32\,\text{eV}$. The reason is as follows. Hydrogen can be ionised by multiple absorption of low-energy photons from the $2S$ or $2P$ exited states. Later recombination by a cascade passing through the $2P$ state can produce a photon of the correct energy (Lyman $\alpha$ line), which can itself excite another atom into the same excited state, which in turn can be ionised by abundant low-energy photons. As photons from the $2P \rightarrow 1S$ transition are confined in the universe, recombination is not possible via a direct cascade through the $2P$ level. The only leakage of the Lyman-$\alpha$ photons passes through the two-photon decay of the $2S$ state. The lifetime of this state is $\approx 0.1$ s; therefore, hydrogen recombination is a nonequilibrium process. The rate of loss of free electrons in the plasma, $n_e$, is given by

$$\frac{dn_e}{dt} = -R(n_e)^2 \frac{\Lambda_{2\gamma}}{\Lambda_{2\gamma} + \Lambda_U(T)} \tag{17.15}$$

where $R$ is the recombination coefficient, $\Lambda_{2\gamma}$ is the two-photon decay rate and $\Lambda_U(T)$ is the stimulated upwards transition rate from the $2S$ state. The time and the temperature of the decoupling are determined by the $2S \rightarrow 1S$ leakage given by (17.15) more correctly than by equilibrium thermodynamics. Even though the analysis can be simplified by assuming that the hydrogen atom has just two levels ($1S$ and $2S$), other parameters strongly depend on the temperature and densities and it is better to use the results of computer simulations than to try to put it *on the back of an envelope*.

The resulting temperature is, however, not much lower than the one calculated above. The transition from an opaque to a transparent universe took place at $T \approx 3{,}000\,\text{K}$ and $z_{\text{dec}} \approx 1{,}300$. Although at $z_{\text{dec}}$, the mean free path of photons increased dramatically, photons still interacted with free electrons via Thomson scattering to a significant extent. Therefore, the photon background that we observe comes from the so-called last scattering surface, where the redshift was less than $z \approx 1{,}000$. At present, the decoupled radiation is a perfect black body spectrum, with temperature $2.7\,\text{K}$.

Let us estimate the time when the decoupling took place. Using the time and the temperature for the matter-dominated period from (17.8) ($T \propto t^{-2/3}$), we obtain

$$t_{\text{dec}} \approx 14 \cdot 10^9 \, \text{yrs} \cdot \left(\frac{2.7}{3,000}\right)^{3/2} \approx 400,000 \, \text{yrs} . \tag{17.16}$$

### 17.1.3 Primordial Abundance of the Elements

The composition of the universe, limiting ourselves to normal matter (from hydrogen to uranium), is as follows: 75% hydrogen, 24% helium and only 1% heavier elements. Carbon and heavier elements are formed in stars. According to the big bang model, helium was formed in the last phase of the early universe. The current mass ratio of 1:3 for helium to hydrogen has been barely altered from its primordial value by synthesis in stars.

When the universe was at a temperature $kT \geq (m_n - m_p - m_e)c^2) = \Delta mc^2$ and at a high density, there was a thermal equilibrium between protons and neutrons due to reactions proceeding via the weak interaction,

$$p + e \leftrightarrow n + \nu \tag{17.17}$$
$$p + \bar{\nu} \leftrightarrow n + e^+ . \tag{17.18}$$

This means that the reaction rates of these reactions were sufficiently rapid to maintain the balance despite cooling. When the temperature reached $kT \leq \Delta mc^2$, the balance tilted in favour of protons

$$\frac{n_n}{n_p} = e^{-\Delta mc^2/kT} , \tag{17.19}$$

where the $n$'s denote the number densities of neutrons and protons. The survival of some neutrons is due to the neutron freeze out, already at a temperature $T \approx 1.2 \, \text{MeV}$. Due to the very weak interaction of low-energy neutrinos, the reaction rates of (17.18) is slower than the cooling of the universe. We will not try to estimate this temperature here. At this time, the neutron fraction of the total baryon number (17.19) was $\approx 34\%$. Further neutron decays were significantly slower due to the neutron lifetime of $\tau = 14.8$ min.

The synthesis of deuterium and beyond to $^4$He is, though, first possible at $kT_D \approx 0.066 \, \text{MeV}$. The reaction

$$p + n \rightarrow D + \gamma \tag{17.20}$$

has a Q value of 2.23 MeV.

Let us estimate the temperature at which deuterons become stable against gamma disintegration. Again, as in the case of $3\alpha$ fusion (Chap. 15) and hydrogen recombination (Sect. 17.1.2), we equate the chemical potentials before and after the reaction. We can use the formulae of the previous Sect. 17.1.2 just by replacing $n_p$, $n_e$, $n_H$ by $n_p$, $n_n$, $n_D$, the hydrogen spin factor 4 by the deuteron spin factor 3, and the deuteron binding by the hydrogen binding.

$$\frac{n_{\mathrm{p}}n_{\mathrm{n}}}{n_{\mathrm{D}}} = \frac{4}{3}\left(\frac{m_{\mathrm{p}}m_{\mathrm{n}}kT}{2\pi\hbar^2 m_{\mathrm{D}}}\right)^{3/2}\mathrm{e}^{-Q/kT}. \tag{17.21}$$

We have again assumed that the cross-over took place at $n_{\mathrm{n}} \approx n_{\mathrm{D}}$. The proton density, $n_p$, can be obtained by extrapolating its present density, $n_{\mathrm{p}}(t_0) = 0.15\ \mathrm{m}^{-3}$, to the temperature $T_{\mathrm{D}}$ when fusion to form deuterium began,

$$n_{\mathrm{p}} \approx n_{\mathrm{p}}(t_0)\left(\frac{T_{\mathrm{D}}}{2.73\ \mathrm{K}}\right)^3. \tag{17.22}$$

As a trick, we have made reference to the cosmic background radiation, which has an accurate temperature 2.73 K at the well-defined present time, $t_0$. Fortunately, the cubic dependence of density on temperature is valid both in the matter- and radiation-dominated eras.

The result is $kT_{\mathrm{D}} = 0.066\,\mathrm{MeV}$ ($T_{\mathrm{D}} \approx 10^{8.88}$ K).

When did deuterium fusion start? At the moment of radiation Decoupling, the universe was 400,000 years old and its temperature was 3,000 K. The temperature in a radiation-dominated system is inversely proportional to the square root of time; therefore, at the temperature of $7.7 \cdot 10^8$ K, which corresponds to $kT_{\mathrm{D}} = 66\,\mathrm{keV}$, the time $t_{\mathrm{D}}$ was

$$t_{\mathrm{D}} = t_{\mathrm{dec}}\left(\frac{T_{\mathrm{dec}}}{T_{\mathrm{D}}}\right)^2 \approx 400,000\left(\frac{3000}{10^{8.88}}\right)^2\ \mathrm{yrs} = 175\mathrm{s} \approx 3\,\mathrm{min}. \tag{17.23}$$

These are the famous "first three minutes", the end of the early universe and the beginning of primordial element synthesis. During the synthesis period, neutrons were decaying, so only 12% of the total number of baryons survived as neutrons inside helium. It is fortunate for nature that the lifetime of the neutron (14 min) is sufficiently longer than the time before the synthesis of deuteron (3 min); otherwise, they would all have decayed and the universe would consist only of protons and electrons.

## 17.2   Some Problems with the Big Bang Model

**For cosmologists**: Nobody likes models that start with a singularity. The expansion of the universe is experimentally well established. But its mechanism is not understood. Even worse, the big bang model needs, at the very beginning, a fast expanding period, the so-called inflationary stage, in order to explain why we only have a very limited horizon of the early universe. What was the mechanism of this expansion? It is rather well established that the universe is flat and that its energy density is at the critical value. However, 70% of this energy is a mysterious dark energy and 27% of it a mysterious dark matter.

**For particle physicists**: How did the fermion–antifermion asymmetry in the universe arise? What is the nature of dark matter? Can gravitation be unified with the other interactions?

### 17.2.1 Particle–Antiparticle Asymmetry

The contemporary abundance of the elements of normal matter relies on three coincidences: the forging of the heavy elements out of helium inside stars, the production of helium in the early universe, and the fermion–antifermion asymmetry, which made the existence of normal matter possible in the first place.

Let us consider the universe shortly after gravitation has decoupled from the other interactions. Radiation dominates this universe, where the particles and antiparticles could annihilate and be produced again through pair creation. As long as the temperature of the universe was sufficiently high, there was an equilibrium between radiation and particle–antiparticle pairs. When, though, due to the cooling of the universe, the radiation ceased to have enough energy to create particles and antiparticles, the fermions available annihilated each other. Because pair creation yielded equal numbers of fermions and antifermions, one would expect that fermionic matter would fully annihilate itself during this cooling down phase. If that had been the case, the universe would now solely consist of background radiation and, perhaps, dark matter and dark energy.

How the matter we observe survived this great annihilation is unclear. The fraction of fermions that survive can be estimated rather well. All cosmological models agree that the number of photons in the universe has not changed significantly during the cooling. After the annihilation, there was only radiation and the surviving electrons and nucleons (protons and those neutrons bound inside $^4$He), which radiation could scatter off. This did not significantly alter the number of photons. About 400,000 years after the big bang, the energy of the radiation was so low that it could not prevent the recombination of protons and $^4$He nuclei with electrons to form neutral atoms. The universe became transparent and radiation separated from matter. This is the origin of the observed cosmic background radiation. The ratio of the numbers of photons and nucleons is around ten billion to one! This implies that the probability of a fermion surviving the annihilation phase was $10^{-10}$.

Three conditions have to be fulfilled to explain the fermion–antifermion asymmetry inside a big bang model: thermal nonequilibrium, CP violation and baryon number violation. Thermal nonequilibrium is easy to imagine in the big bang model. In this phase, the cooling of the universe has to be quicker than the reaction rate, which maintains equilibrium. The CP violation and baryon-number violation, which are responsible for the asymmetry at $T \approx 10^{15}$ GeV, must still be detectable today. The tiny CP violation seen in $K^0$ and $B^0$ systems is, when extrapolated to higher energies, insufficient to produce this asymmetry. The proton also seems to be more stable than grand unified theory would lead us to expect.

There is a possibility, however, that the baryon–antibaryon asymmetry is the consequence of the lepton–antilepton asymmetry produced during the time of the GUT. The difference, $B - L$ (Baryon–Lepton number), is, in most theoretical models, conserved. For the charged leptons, this is obviously necessary if one starts out with charge conservation and an electrically neutral universe. Furthermore, the charged and neutral leptons are related within the same weak doublets.

The lepton sector is experimentally less thoroughly searched for the lepton number and CP violation than the baryon sector. Therefore, it cannot be excluded that the solution of the fermion–antifermion asymmetry is to be found in the lepton sector.

## 17.2.2  Dark Matter

Already in the 1930s, Fritz Zwicky recognised that the relative movements of the galaxies could only be explained if the galaxies and their surroundings contained about five times as much matter as was observed through telescopes. The lensing effects of gravitation on photons are even more spectacular. Tiny shifts in the position of stars, observed when light passes through the gravitational field of the sun during a total solar eclipse, served to confirm general relativity theory. Today, it is possible to observe much larger lensing effects due to the gravitational field of large clusters of galaxies. In these cases, too, the effects observed can only be explained when the galaxy clusters are ascribed around five times as much mass as their visible mass. The name "dark matter" has been given to this material because it neither absorbs nor emits light. It also cannot interact via the strong interaction; otherwise, it would be noticed by a high collision rate with the ordinary matter.

There are various speculations about the nature of dark matter. The most attractive seem to be the idea that it is made of heavy, weakly interacting particles that are relics of the big bang. If that really is the case, then such particles could be easily incorporated into the scenarios of symmetry breaking. Heavy weakly interacting particles would be bound by gravity to galaxies, but because they, unlike normal matter, would not clump together, their extension would be much greater than that of normal matter. The search for such particles is a challenge for experimentalists. Because these particles would only weakly interact, the detector would have to be as large as is required to detect neutrinos. The energy of the particles would be small and their speeds comparable with those of other objects in the Milky Way. The detectable recoil of normal atoms due to a weak interaction collision would produce a few pairs of ions. To distinguish such recoil signals from signals of ionising particles, one would need to simultaneously detect phonons, and so detectors will have to work at temperatures of liquid helium. The first detectors that partially fulfill these requirements are already in action in underground laboratories in Gran Sasso (Italy), Frejus (France) and Soudan (USA).

### 17.2.3  Physics at the Planck Scale

The big bang model treats the universe as a gas of which the constituents are currently galaxies; though, the further back in time one goes, the smaller the constituents were. The dynamics of the expanding universe is described by general relativity theory. It is clear that the model's classical physics is insufficient as $R \rightarrow 0$. Difficulties will appear, at the latest, when the thermal energy of particles is large enough for the de Broglie wavelength to be smaller than the Schwarzschild radius. Quantum black holes clearly cause difficulties with our usual concept of a background space–time. If we equate the de Broglie wavelength, $2\pi\hbar/mc$, and the Schwarzschild radius $2Gm/c^2$ (Sect. 15.3.2), we obtain the characteristic mass of quantum gravitation, the Planck mass and its accompanying length and time scales,

$$m_P = \sqrt{\frac{\hbar c}{G}} \simeq 10^{19} m_p \simeq 10^{19}\,\mathrm{GeV}/c^2 \tag{17.24}$$

$$l_P = \frac{\hbar c}{m_P c^2} \simeq 10^{-35}\,\mathrm{m} \tag{17.25}$$

$$t_P = \frac{l_P}{c} \simeq 10^{-43}\,\mathrm{s}\,. \tag{17.26}$$

Here, we have used the value $G = 10^{-38}\hbar c/m_p^2$ for the gravitational constant (see (15.8)).

An additional motivation for introducing the Planck scale follows the striving to unify all interactions. The fact that there is still no experimental support for the grand unified theory, should not stop us from speculating about a further possible unification of the other interactions with gravitation. The Planck scale, at which this ultimate unification could take place, can be defined in terms of the mass at which the gravitational coupling constant also reaches a value comparable with the other coupling constants, i.e.,

$$\alpha_G = \frac{Gm_P^2}{\hbar c} = \frac{G(E_P/c^2)^2}{\hbar c} \sim 1\,. \tag{17.27}$$

This definition of the Planck scale is an alternative to the above considerations of the de Broglie wavelength and yields the same result.

The Planck time marks the beginning of the classical phase of the big bang.

Gravitation is the dominant interaction in the universe. It is reasonable to ask whether it has left any traces from the period when it was unified with the other interactions. The existence of the three particle families could be such a trace. The three families have identical properties as far as the strong, electromagnetic and weak interactions are concerned. They differ in their masses – only gravitation distinguishes between the different families!

# Literature

G. Börner, *The Early Universe* (Springer, Berlin, 2003)

J.A. Peacock, *Cosmological Physics* (Cambridge University Press, Cambridge, 1999)

D. Perkins, *Particle Astrophysics* (Oxford University Press, Oxford, 2003)

M. Treichel, *Teilchenphysik und Kosmologie* (Springer, Berlin, 2000)

S. Weinberg, *The First Three Minutes* (Basic Books, Inc., Publishers, New York, 1977)

> *Nous pardonnons souvent à ceux qui nous ennuient, mais nous ne pouvons pardonner à ceux que nous ennuyons.*
>
> La Rochefoucauld

# Physical Constants

| | | | |
|---|---|---|---|
| Speed of light | $c$ | $2.99792458 \cdot 10^8$ | $\mathrm{m\,s^{-1}}$ |
| | $h$ | $6.626070040(81) \cdot 10^{-34}$ | J s |
| | $\hbar = h/2\pi$ | $1.054571800(13) \cdot 10^{-34}$ | J s |
| Planck's constant | | $6.582119514(40) \cdot 10^{-22}$ | MeV s |
| | $\hbar c$ | $197.3269788(12)$ | MeV fm |
| | $(\hbar c)^2$ | $0.3893793656(48)$ | $\mathrm{GeV^2\,mb}$ |
| Atomic mass unit | $u = M_{^{12}C}/12$ | $931.4940954(57)$ | $\mathrm{MeV}/c^2$ |
| Proton mass | $m_\mathrm{p}$ | $938.2720813(58)$ | $\mathrm{MeV}/c^2$ |
| Neutron mass | $m_\mathrm{n}$ | $939.56536(8)$ | $\mathrm{MeV}/c^2$ |
| Electron mass | $m_\mathrm{e}$ | $0.5109989461(31)$ | $\mathrm{MeV}/c^2$ |
| Elementary charge | $e$ | $1.6021766208(98) \cdot 10^{-19}$ | A s |
| Dielectric constant | $\varepsilon_0 = 1/\mu_0 c^2$ | $8.854187817 \cdot 10^{-12}$ | $\mathrm{A\,s\,V^{-1}\,m^{-1}}$ |
| Permeability of vacuum | $\mu_0$ | $4\pi \cdot 10^{-7}$ | $\mathrm{V\,s\,A^{-1}\,m^{-1}}$ |
| Fine structure constant | $\alpha = e^2/4\pi\varepsilon_0\hbar c$ | $1/137.035999139(31)$ | |
| Classical electron radius | $r_\mathrm{e} = e^2/4\pi\varepsilon_0 m_\mathrm{e} c^2$ | $2.8179403227(19) \cdot 10^{-15}$ | m |
| Compton wavelength | $\lambdabar_\mathrm{e} = r_\mathrm{e}/\alpha$ | $3.8615926764(18) \cdot 10^{-13}$ | m |
| Bohr radius | $a_0 = r_\mathrm{e}/\alpha^2$ | $0.52917721067(12) \cdot 10^{-10}$ | m |
| Bohr magneton | $\mu_\mathrm{B} = e\hbar/2m_\mathrm{e}$ | $5.7883818012(26) \cdot 10^{-11}$ | $\mathrm{MeV\,T^{-1}}$ |
| Nuclear magneton | $\mu_\mathrm{N} = e\hbar/2m_\mathrm{p}$ | $3.1524512550(15) \cdot 10^{-14}$ | $\mathrm{MeV\,T^{-1}}$ |
| | $\mu_\mathrm{e}$ | $1.001159652187(4)$ | $\mu_\mathrm{B}$ |
| Magnetic moments | $\mu_\mathrm{p}$ | $2.792847351(28)$ | $\mu_\mathrm{N}$ |
| | $\mu_\mathrm{n}$ | $-1.9130427(5)$ | $\mu_\mathrm{N}$ |
| Avogadro's number | $N_\mathrm{A}$ | $6.022140857(74) \cdot 10^{23}$ | $\mathrm{mol^{-1}}$ |
| | $k$ | $1.38064852(79) \cdot 10^{-23}$ | $\mathrm{J\,K^{-1}}$ |
| Boltzmann's constant | | $8.617343(15) \cdot 10^{-5}$ | $\mathrm{eV\,K^{-1}}$ |
| | $G_\mathrm{N}$ | $6.67408(31) \cdot 10^{-11}$ | $\mathrm{N\,m^2\,kg^{-2}}$ |
| Gravitational constant | $G_\mathrm{N}/\hbar c$ | $6.70861(31) \cdot 10^{-39}$ | $c^4/\mathrm{GeV^2}$ |
| Fermi constant | $G_\mathrm{F}/(\hbar c)^3$ | $1.1663787(6) \cdot 10^{-5}$ | $\mathrm{GeV^{-2}}$ |
| Weinberg angle | $\sin^2\theta_\mathrm{W}$ | $0.23129(5)$ | |
| Mass of the $W^\pm$-Boson | $M_\mathrm{W}$ | $80.385(15)$ | $\mathrm{GeV\,c^{-2}}$ |
| Mass of the $Z^0$-Boson | $M_\mathrm{Z}$ | $91.1876(21)$ | $\mathrm{GeV\,c^{-2}}$ |
| Strong coupling constant | $\alpha_\mathrm{s}(M_\mathrm{Z})$ | $0.1182(12)$ | |

© Springer-Verlag GmbH Germany 2017
B. Povh and M. Rosina, *Scattering and Structures*,
Graduate Texts in Physics, DOI 10.1007/978-3-662-54515-7

# Literature

C. Patrignani et al., (Particle Data Group). The review of particle physics. Chin. Phys. C **40**, 100001 (2016)

# Index

## A
Age of the universe, 209
$\alpha$-Helix, 87
Antiferromagnetism, 64
Antineutrinos, 31
Antiquarks, 31
Argon, 9
Asteroids, 167, 182, 184
  Ceres, 184
Atomic radius, 8, 59

## B
Baryons, 135
$\beta$ decay, 194
Beta pleated sheet $\beta$-Pleated sheet, 87
Big bang, 207, 208, 214
Binding, 119
Binding energy
  of atoms, 58, 67
  of metals, 120
  of nuclei, 156, 157
  of the helium atom, 56
  of the hydrogen atom, 42
Black holes, 180
Bohr radius, 21, 42, 59
Bond
  covalent, 67, 71, 83
    (2s, 2p) shells, 71
  hydrogen bridge, 82, 84, 85, 87
  ionic, 67
  metallic, 67, 121
  peptide, 85
Bond energy
  molecular, 71, 74
Bose condensate, 101, 113, 123, 126
Bose–Einstein condensation, 104

occupation number, 106
Bose–Einstein statistics, 97
Bose liquid, 90
Boson
  Higgs, 201
  W, 23, 185–187, 196
  weak, 199
  $Z^0$, 22, 196
Bosonic gas, 104, 106, 113
Bosonic liquids, 111, 113
Bottomium, 134, 135
Bragg scattering, 89
Bremsstrahlung, 26
  of gluons, 31, 33, 35
  of photons, 35
  spectrum, 28
Brown dwarfs, 172

## C
Cabibbo, 186, 188
Cabibbo–Kobayashi–Maskawa matrix, 187
Carbon, 71
  diamond, 72
  fullerene, 73
  graphene, 72
  graphite, 73
  nanotubes, 73
  syntesis, 177
Casimir effect, 79
Casimir force, 81
Charge distribution, 20, 21
Charge radius, 20
Charmonium, 133, 135
Chiral symmetry breaking, 104
Chirality, 138, 188
Cobalt, 63

© Springer-Verlag GmbH Germany 2017
B. Povh and M. Rosina, *Scattering and Structures*,
Graduate Texts in Physics, DOI 10.1007/978-3-662-54515-7

Printed in the United States
By Bookmasters